*S.Chand's IIT Foundation Series*

## A Compact and Comprehensiv

# IIT Foundatio
# Science

(Physics and Chemistry)

**CLASS – VI**

*S.Chand's IIT Foundation Series*

## A Compact and Comprehensive Book of

# IIT Foundation Science

## (Physics and Chemistry)

## CLASS – VI

S.K. GUPTA
ANUBHUTI GANGAL

Eurasia Publishing House
(An imprint of S. Chand Publishing)

**EURASIA PUBLISHING HOUSE (EPH)**
**(An imprint of S. Chand Publishing)**
A Division of S. Chand & Co. Pvt. Ltd.
7361, Ram Nagar, Qutab Road, New Delhi-110055
**Phone:** 23672080-81-82, 9899107446, 9911310888; **Fax:** 91-11-23677446
www.schandpublishing.com; **e-mail:** helpdesk@schandpublishing.com

**Branches :**

Ahmedabad : Ph: 27541965, 27542369, ahmedabad@schandgroup.com
Bengaluru : Ph: 22268048, 22354008, bangalore@schandgroup.com
Bhopal : Ph: 4274723, 4209587, bhopal@schandgroup.com
Chandigarh : Ph: 2725443, 2725446, chandigarh@schandgroup.com
Chennai : Ph. 28410027, 28410058, chennai@schandgroup.com
Coimbatore : Ph: 2323620, 4217136, coimbatore@schandgroup.com (Marketing Office)
Cuttack : Ph: 2332580; 2332581, cuttack@schandgroup.com
Dehradun : Ph: 2711101, 2710861, dehradun@schandgroup.com
Guwahati : Ph: 2738811, 2735640, guwahati@schandgroup.com
Haldwani : Mob. 09452294584 (Marketing Office)
Hyderabad : Ph: 27550194, 27550195, hyderabad@schandgroup.com
Jaipur : Ph: 2219175, 2219176, jaipur@schandgroup.com
Jalandhar : Ph: 2401630, 5000630, jalandhar@schandgroup.com
Kochi : Ph: 2378740, 2378207-08, cochin@schandgroup.com
Kolkata : Ph: 22367459, 22373914, kolkata@schandgroup.com
Lucknow : Ph: 4076971, 4026791, 4065646, 4027188, lucknow@schandgroup.com
Mumbai : Ph: 22690881, 22610885, mumbai@schandgroup.com
Nagpur : Ph: 2720523, 2777666, nagpur@schandgroup.com
Patna : Ph: 2300489, 2302100, patna@schandgroup.com
Pune : Ph: 64017298, pune@schandgroup.com
Raipur : Ph: 2443142, Mb. : 09981200834, raipur@schandgroup.com (Marketing Office)
Ranchi : Ph: 2361178, Mob. 09430246440, ranchi@schandgroup.com
Siliguri : Ph. 2520750, siliguri@schandgroup.com (Marketing Office)
Visakhapatnam : Ph. 2782609 (M) 09440100555, visakhapatnam@schandgroup.com (Marketing Office)



*First Edition 2012*

*Reprints 2013, 2014, 2015*

**ISBN :** 978-81-219-3906-5 **Code :** 1016 388

PRINTED IN INDIA
By Vikas Publishing House Pvt. Ltd., Plot 20/4, Site-IV, Industrial Area Sahibabad, Ghaziabad-201010
and Published by S. Chand & Company Pvt. Ltd., 7361, Ram Nagar, New Delhi-110055.

# PREFACE AND A NOTE FOR THE STUDENTS

***ARE YOU ASPIRING TO BECOME AN ENGINEER AND BECOME AN IIT SCHOLAR ?***

Here is the book especially designed to motivate you, to sharpen your intellect and develop the right attitude and aptitude and lay solid foundation for your success in various entrance examinations like **IIT, AIEEE, EAMCET, WBJEE, MPPET, SCRA, Kerala PET, OJEE, Raj PET, AMU,** etc.

**SALIENT FEATURES**

1. Content based on the curriculum of the classes for ***CBSE, ICSE, Andhra Pradesh and Boards of School Education of other states.***
2. Full and comprehensive coverage of all the topics.
3. Detailed synopsis of each chapter at the beginning in the form of ***'Key Concepts'***. This will not only facilitate thorough ***'Revision'*** and ***'Recall'*** of every topic but also greatly help the students in understanding and mastering the concepts besides providing a ***back-up*** to classroom teaching.
4. The books are enriched with an exhaustive range of hundreds of thought provoking objective questions in the form of solved examples and practice questions in Question Banks which not only offer a great variety and reflect the modern trends but also invite, explore, develop and put to test the ***thinking, analysing and problem solving skills of the students***.
5. **Answers, Hints** and **Solutions** have been provided to boost up the morale and increase the confidence level.
6. **Self Assessment Sheets** have been given at the end of each chapter to help the students to assess and evaluate their understanding of the concepts and learn to attack the problems independently.
7. Concept maps and crossword puzzles have been given to help the students gain mastery over the fundamentals of each topic.

We hope the series will be able to fulfil its aims and objectives and will be found immensely useful by the students aspiring to become topclass engineers.

Suggestions for improvement and also the feedback from various sources would be most welcome and gratefully acknowledged.

**NOTE**

The text-matter has been thoroughly checked and an earnest effort has been made to make the book error-free.

**October, 2012** **AUTHORS**

# CONTENTS

## Topics in Physics



## Topics in Chemistry

# TOPICS IN PHYSICS

- *Measurement and Motion*
- *Force and Pressure*
- *Work and Energy*
- *Light*
- *Electricity*
- *Magnetism*

Chapter 1

# MEASUREMENT AND MOTION

## Section-A

## MEASUREMENT

### KEY FACTS

1. **A Basic or Fundamental Quantity** is that which cannot be derived from another. Length, mass, time, electric current, temperature are five basic quantities.*
2. **A derived quantity** is that which can be obtained by combining of basic quantities using multiplication and division. Area, volume, density, velocity, force, etc., are derived quantities.
3. **SI Units.** The scientists, all over the world, have accepted a basic set of units for the measurement of physical quantities. This is called the S.I. system of units. The following table shows a list of physical quantities and their respective standard units.

| Physical quantity | SI Unit | |
|---|---|---|
| | Name | Symbol |
| Length | metre | m |
| Area | square metre | $m^2$ |
| Volume | cubic metre | $m^3$ |
| Mass | kilogram | kg |
| Time | second | s |
| Temperature | kelvin | K |
| Current | ampere | A |

4. **Prefixes used in S.I. Units**

A.

| Name | Symbol | Submultiple | Example (1 metre is written as 1 m) |
|---|---|---|---|
| deci | d | $\frac{1}{10} = 10^{-1} = 0.1$ | decimetre (dm) = $\frac{1}{10}$ metre = $\frac{1}{10}$ m |
| centi | c | $\frac{1}{100} = 10^{-2} = 0.01$ | centimetre (cm) = $\frac{1}{100}$ m |
| milli | m | $\frac{1}{1000} = 10^{-3} = 0.001$ | millimetre (mm) = $\frac{1}{1000}$ m |

* The remaining two basic quantities are *luminous intensity* and *amount of substance* with which we are not dealing in this book.

B.

| Name | Symbol | Multiple | Example |
|---|---|---|---|
| deca | da | 10 (10 times) | decametre (dam) = 10 m |
| hecto | h | 100 (100 times) | hectometre (hm) = 100 m |
| kilo | k | 1000 (1000 times) | kilometre (km) = 1000 m |
| mega | M | 1000000 (10 lakh times) | megametre(Mm) = 1000000 m |

5. The following units are used to measure objects, like very small bacteria, insects, viruses etc., micro (μ, pronounced as mu), nanometre (nm), Angstrom (Å) and picometre (pm).

**1 μm (micron or micrometre)** = 1 millionth of a metre = $\frac{1}{1000000}$ m = $10^{-6}$ m

**1 nm (nanometre)** = $10^{-9}$ m = one billionth of a metre = $\frac{1}{1000000000}$ m = $10^{-9}$ m

1 Å (Angstrom) = $10^{-10}$ m

1 pm (picometre) = $10^{-12}$ m.

6. The following units are used to measure very large distances like the distances between stars and planets etc.

**1 light year**= 9.465 × $10^{15}$ m

= 9.46 trillion km

= **About 9.5 lakh crore metre**

= 9,500,000,000,000 km

**1 Parsec** = **3.2616 light years**

1 light year is the distance a ray of light can travel in one year. If a star is 5 light years away from us, it means that this distance is equal to the distance travelled by light in 5 years. Thus the light from the star that reaches us today started from the star 5 year ago.

## Question Bank-1(a)

1. Arrange the following lengths in the increasing magnitude.

   1 metre, 1 megametre, 1 centimetre, 1 kilometre, 1 millimetre, 1 micrometre.

2. If a woman has a mass of 45,000,000 mg, what is her mass in grams and kilograms?

3. Which SI units would you use for the following measurements?

   (a) the length of a swimming pool

   (b) the mass of the water in the pool

   (c) the time it takes a swimmer to swim a lap

4. Which of the following is the best estimate in metres of the height of a mountain?

   (a) 1 m (b) 100 m (c) 1 km (d) 1 Mm

5. Ten metres is equal to

   (a) 100 cm (b) 1,00,000 mm

   (c) 1,000 cm (d) 1,000 μm

6. A certain bacterial cell has a diameter of 0.50 μm. The tip of a pin is about 1100 μm in diameter. How many of these bacterial cells would fit on the top of the pin?

7. **List an appropriate SI base unit (with a prefix as needed) for measuring the following.**

   (a) the time it takes to play a CD in your stereo

   (b) the mass of a SUV

   (c) the length of a soccer field

   (d) the diameter of a large pizza

(e) the distance between New Delhi and Jaipur

(f) your mass

(g) the length of your school auditorium

(h) your height

**8. Estimate the magnitude of the lengths in metres of each of the following:**

(a) a ladybug (b) your leg

(c) your school building (d) a giraffe

**9. Express each of the following as indicated:**

(a) 3.5 dm expressed in mm

(b) 3 h 20 min expressed in seconds

(c) 0.59 km expressed in centimetres

(d) 380 μm in centimetres

(e) 0.592 mg expressed in grams

(f) 25 g expressed in micrograms

(g) 36 km/h expressed in metres per second

**10.** What is the SI base unit for length?

(a) inch (b) foot

(c) metre (d) kilometre

**11.** A light year (*ly*) is a unit of distance defined as the distance light travels in one year. Numerically, 1 *ly* = 9 500 000 000 000 km. How many metres are in a light year?

(a) $9.5 \times 10^{10}$ m (b) $9.5 \times 10^{12}$ m

(c) $9.5 \times 10^{15}$ m (d) $9.5 \times 10^{18}$ m

**12.** Ankit is measuring how fast bacteria grow in a dish by measuring the area that the bacteria cover. On day 1, the bacteria cover 0.35 cm². On day 2, they cover 0.70 cm². On day 3, they cover 1.40 cm². What is the best prediction for the area covered on day 4?

(a) 1.50 cm² (b) 3 cm²

(c) 2.80 cm² (d) 2.90 cm².

**13. Create a concept map using the following words:**

*Measurement, SI units, Length, Mass, Time, Area, Volume, CGS units, Large units, Very small unit, megametre, micron, light year, metre, kilogram, cubic metre, centimetre, gram, cubic centimetre, square metre, square centimetre*

**14. Solve the following crossword with the help of the given clues.**

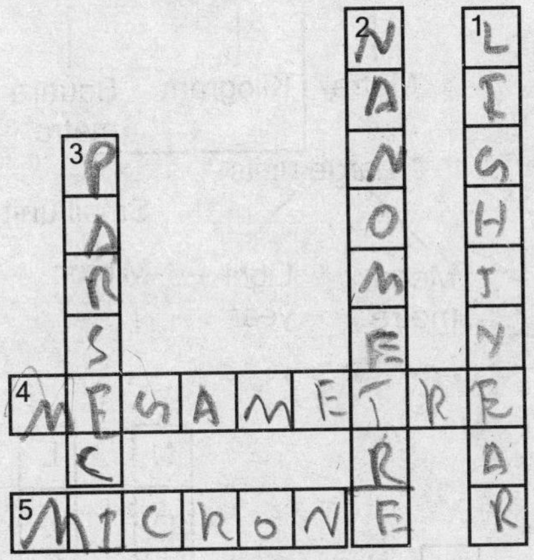


**ACROSS**

4. 1 million (ten lakh) metres

5. one- millionth $\left(\frac{1}{10 \text{ lakh}}\right)$ of a metre

**DOWN**

1. Distance light travels in one year.

2. One billionth $\left(\frac{1}{100 \text{ crore}}\right)$ of a metre

3. A unit of distance used in astronomy. About $3\frac{1}{4}$ light years.

## Answers

**1.** micrometre, millimetre, centimetre, metre, kilometre, megametre.

**2.** 45,000 g; 45 kg

**3.** (a) metres (b) kilograms

(c) seconds

**4.** (c) **5.** (c)

**6.** 2200 bacterial cells

**7.** (a) s or das (b) kg (c) m
(d) cm or dm (e) km (f) kg
(g) m (h) cm, dm, or m.

**8.** (a) 1 cm = 0.01 m (b) 1 m
(c) 10 m or more (d) 10 m

**9.** (a) 350 mm (b) 12000 s (c) 59000 cm
(d) 0.038 cm (e) 0.000592 g
(f) $25 \times 10^6$ μg (g) 10 m/s

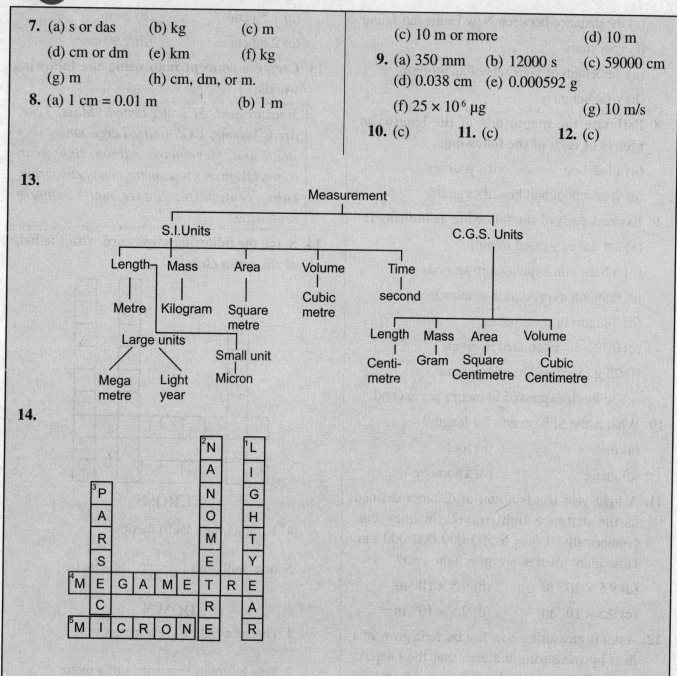

**10.** (c) **11.** (c) **12.** (c)

**13.**

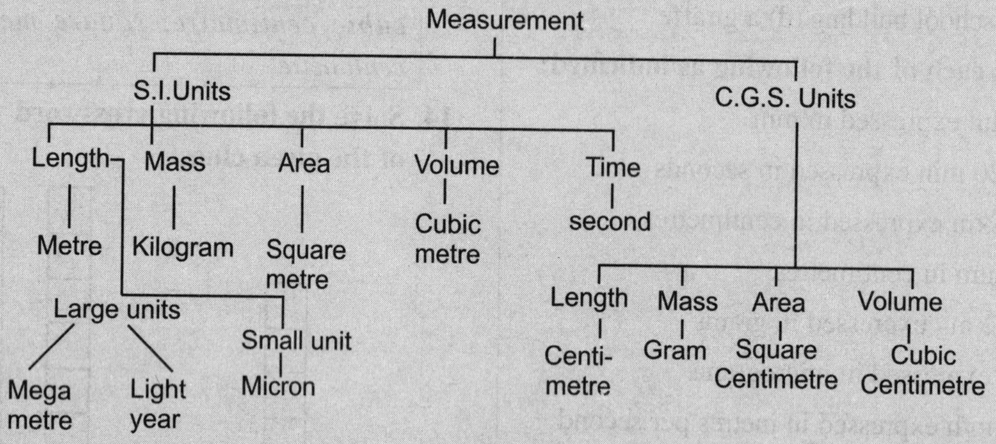


**14.**

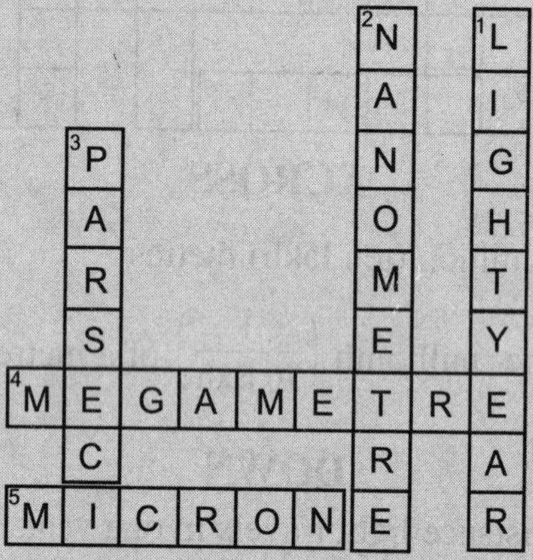

# Section-B
# MOTION

## KEY FACTS

**I.** An object is said to be in motion if it changes its position with respect to other objects around it, when a force acts on it.

## II. Different types of motion

1. **Translatory motion.** The type of motion in which the entire position of the object changes, that is, all its parts move simultaneously through the same distance is called translatory motion.

   **Ex.** Motion of a car, a bird, an aeroplane. All of them move as a unit.

   A translatory motion may be of two types — a straight line motion called **Rectilinear Motion** or a curved line motion called **Curvilinear Motion**.

   (i) **Rectilinear Motion**–Object moves in a straight line, *e.g.*, a car travelling along a straight road.

   (ii) **Curvilinear Motion–**When objects moving along a straight line take a curve.

   **Ex.** A car moving along a curve on the road.

2. **Circular Motion**–Object moves along a circular path and the distance of the object from the central point remains the same.

   **Ex.** (i) Movement of tip of the hand of a clock or watch on the dial.

   (ii) Motion of the blades of a ceiling fan when it is switched on.

3. **Rotational Motion**–When an object turns (or spins) about a fixed axis, it is called rotational motion.

   **Ex.** Spinning top, spinning of earth on its axis.

4. **Revolution**– An object moving around a circular path repeatedly.

   **Ex.** Blades of a moving fan, motion of handle of a sewing machine, Earth revolving round the sun.

5. **Oscillatory Motion**–An object passing through the same point while moving back and forth between two points.

   **Ex.** Motion of a swing, motion of the needle end of a sewing machine.

6. **Vibratory Motions–**Oscillatory motions which begin very fast and then slow down soon to position of rest. Vibrating objects produce sound.

   **Ex.** Plucked string of a sitar.

7. **Periodic Motion–**The motion which repeats itself after regular intervals of time.

   **Ex.** The motion of seconds' hand of a watch, the revolution of earth around the sun, swinging of a pendulum, Motion of pendulum of a "pendulum clock".

8. **Non-Periodic Motion–**Sudden movement without any regularity.

   **Ex.** Earthquake, storm, landslide.

9. **Mixed Motion**–More than one type of motion at the same time, *i.e.*, a combination of two or more motions.

   **Ex.** (i) Motion of earth $\rightarrow$ circular motion, periodic motion and rotational motion.

   (ii) A ball rolling on the gound $\rightarrow$ rotational motion and rectilinear motion.

## Question Bank-1(b)

**1.** What are the similarities and differences between the motion of a bicycle and a ceiling fan that has been switched on.

**2. Fill in the blanks.**

(i) Motion of a needle on a sewing machine is ________ motion.

(ii) Motion of a wheel on a bicycle is ________ motion.

(iii) The motion of a plucked string of a violin or sitar is ________ motion.

(iv) A bird flying in the sky posseses ________ motion.

**3.** Motion of a screw while going into the wood is an example of

(a) linear and spin motion

(b) rotation and revolution

(c) rotation and spin motion

(d) rotation and linear motion

**4.** Motion of pendulum of a clock is an example of

(a) rotational motion (b) curvilinear motion

(c) rectilinear motion (d) periodic motion

**5.** Name the type of motion seen in the following:

**Word Bank**: *revolution, periodic motion, rotation, oscillatory motion, curvilinear motion, rectilinear motion.*

(a) Earth rotating on its axis

(b) Blades of a moving fan

(c) Needle end of a sewing machine

(d) A rocket fired into space

(e) An apple falling from a tree

(f) A car moving along a road

(g) A car moving along a curve on the road

(h) Motion of the branch of a tree moving to and fro.

**6.** Motion of earth has

(a) circular motion (b) periodic motion

(c) rotational motion (d) All the three types

**7. Match the following:**

| | |
|---|---|
| (i) A buzzing bee | (a) Time taken by the bob to complete one oscillation |
| (ii) A bullet fired from a gun | (b) Rotatory |
| (iii) Guitar string | (c) Vibratory |
| (iv) Time period of a simple pendulum | (d) Periodic |
| (v) Heart beat | (e) Random motion |
| (vi) Potter's wheel | (f) Linear and rotatory |
| (vii) A cricket ball bowled | (g) Linear |
| (viii) A flying kite | (h) Curvilinear |

**8.** Which part of the moving cycle undergoes rotatory motion ?

(a) A

(b) B

(c) C

(d) All of these

**9.** Praveen is drilling a hole in the wall. What type of motion is caused ?

(a) rotatory

(b) translatory

(c) curvilinear

(d) None

**10.** The motion of sea waves is

(a) rectilinear (b) curvilinear

(c) oscillatory (d) both a and c

**11.** When you play soccer, the motion described by the football is

(a) curvilinear

(b) circular

(c) oscillatory

(d) non-uniform

**12.** Which of the following objects does not have more than one type of motion ?

(a) screw (b) rolling ball

(c) scooter's wheel (d) child on a seesaw

**13.** A cork is placed on the surface of water. A small stone is dropped in the water. As a result wave motion is produced on the surface of water and the cork starts moving. What kind of motion does the cork describe ?

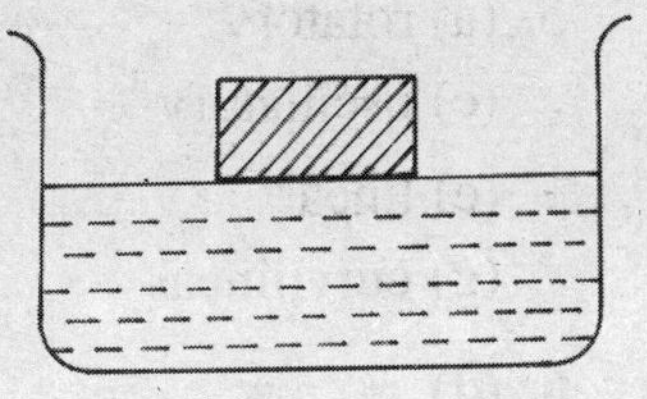

(a) periodic

(b) linear

(c) circular

(d) both periodic and circular

**14. Using the following words draw a concept map.**

**Words** : *Motion, Translatory, Circular, Rotatory, Oscillatory, Repetitive, Periodic, Non-periodic, Vibratory, Rectilinear, Curvilinear.*

**15. Solve the following crossword with the help of the given clues :**

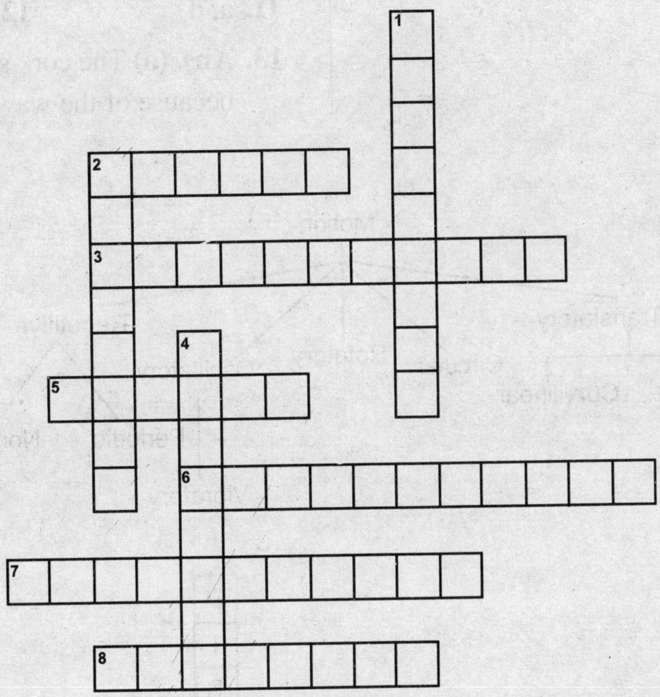


**ACROSS**

**2.** Irregular motion such as motion of a ball during a game of hockey or football.

**3.** The motion in which all parts of the body travel through the same distance

**5.** Movement of a body

**6.** Motion of a body along a curved path.

**7.** Motion of a body like a pendulum

**8.** Occuring or appearing at intervals

**DOWN**

**1.** Motion of a plucked string of a sitar

**2.** Spinning of a body about a fixed axis

**4.** Motion of a girl sitting on a merry–go–round

## Answers

1. *Similarities:* Wheel of a bicycle and ceiling fan (when on) both show circular motion

   *Differences:* Cycle moves in rectilinear motion while a ceiling fan does not have this kind of motion.

2. (i) Oscillatory

   (ii) Both the circular (rotatory) and linear motion

   (iii) Vibratory

   (iv) Random

3. (d)

4. (d)

5. (a) rotatory (b) circular\rotatory

   (c) oscillatory (d) linear

   (e) linear (f) linear

   (g) curvilinear (h) oscillatory

6. (d)

7. (i) –e, (ii) –f, (iii) –c, (iv) –a,

   (v) –d, (vi) –b, (vii) –h, (viii) –e

**8.** (ii) **9.** a, b **10.** d

**11.** a, d **12.** d

**13. Ans. (a)** The cork starts moving up and down because of the waves on the surface of water.

**14.**

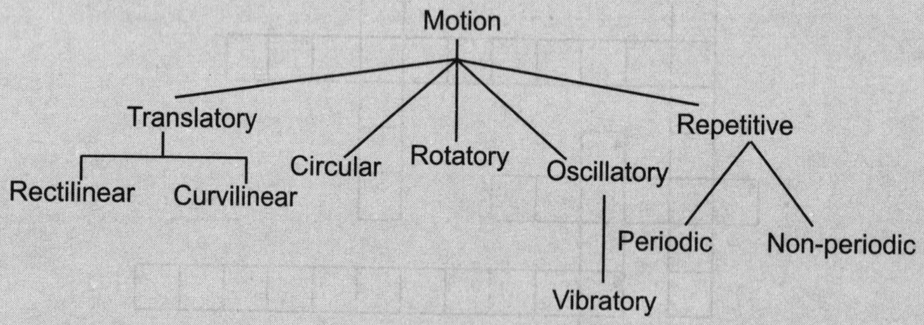


**15.**

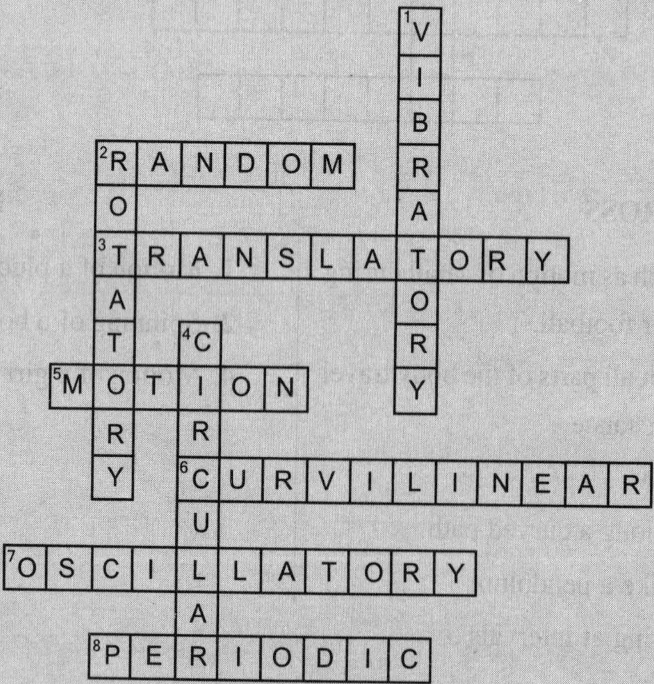

## Self Assessment Sheet-1(a)

**1. Match the following**

| **Column A** | **Column B** |
|---|---|
| (i) dm | a. one hundredth $\left(\frac{1}{100}\right)$ part |
| (ii) mg | b. Megametre |
| (iii) kilo | c. One thousand (1000) |
| (iv) centi | d. milligram |
| (v) Mm | e. decimetre |

**2.** One dozen coins were arranged one above the other. Their total height was 6 cm 6mm. The thickness of each coin is

(a) 6.4 mm (b) 6.6 mm

(c) 5.5 mm (d) None of these

**3.** Arrange the following symbols in the increasing order of lengths they represent.

dm cm m km dam Mm mm μm

**4.** If a tunnel is dug along the diameter of the earth and a ball is dropped into the tunnel, it will have

(a) linear motion (b) circular motion

(c) oscillatory motion (d) translatory motion

**5.** Which of these represents one complete oscillation of the pendulum shown here.

(a) PQR

(b) PQQP

(c) QRRP

(d) PQR RQP

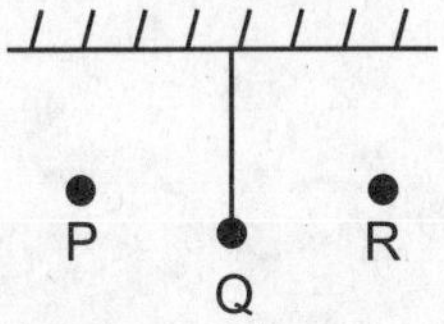


**6.** A satellite is orbiting the earth in such a manner that the satellite is always straight above India. It is at a height of about 36,000 km. Which of the following is true ?

(a) Its period of rotation is 24 hours.

(b) Its period of revolution is 24 hours.

(c) Its period of rotation is 48 hours.

(d) Its period of revolution is 48 hours.

**7. Answer true or false :**

(i) The SI unit of length is cm.

(ii) The motion of moon around the earth is circular.

(iii) The motion of the ball in a game of football is curvilinear.

(iv) The thickness of 80 turns of a wire is found to be 72 cm. The thickness of the wire is 9 cm.

(v) The motion of the seconds hand of a clock is rotational.

**8.** The motion of the arms of soldiers taking part in march past.

(a) circular

(b) oscillatory

(c) rotatory

(d) non-periodic

**9.** What is common to the motion exhibited in the following pictures.

Blades of an exhaust fan

A child on a merry go round

The motion of a spinning wheel

A couple taking 7 rounds in marriage

(a) All motions are translatory

(b) All motions are curvilinear

(c) All motions are rotatory

(d) All motions are circular

**10. Which one is odd man out ?**

(a) A car taking turn on a curved road

(b) Motion of a swing

(c) Motion of needle end of a sewing machine

(d) Motion of an engine piston

## Answers

1. (i) –e, (ii) –d (iii) –c (iv) –a (v) –b
2. (c)
3. µm , mm, cm, dm, m, dam, km, Mm
4. (c) 5. (d) 6. (b)
7. (i) False, the SI unit of length is metre
   (ii) True
   (iii) False. It is random motion
   (iv) False. It is 0.9 cm
   (v) True
8. (b)
9. All of them are examples of circular motion.
10. (a). It is curvilinear motion. The others are periodic motions.

Chapter 2

# FORCE AND PRESSURE

## KEY FACTS

1. **Force** is a push or pull which changes or tends ***to change the state of rest or motion of the body, i.e., the initial state of the body***.
2. The standard unit of force is **newton** (N), named after the great English Scientist, Sir Isaac Newton.
3. Objects thrown drop to the ground unless something holds them up. This change in position is affected by **gravity. Gravity** is a pull towards the centre of the Earth.
4. **Friction** is a force that makes an object slow down when it rubs against another object.
   (i) Football players wear special shoes designed to increase the friction that occurs between their feet and artificial turf. This helps players' feet grip the playing field and reduces the likelihood of slipping.
   (ii) Skiers, cyclists and speed skaters crouch down to reduce the amount of friction that occurs between their bodies and the wind.

**Types of forces**

The forces are classified into two classes.

(i) **Contact forces** which require physical contact between the objects, *e.g.*, push, pull, twist, friction, etc.

(ii) **Non-contact forces** which do not require physical contact between two bodies such as gravitational force, magnetic force and electrostatic force.

5. The force exerted by an electrically charged object is called **electrostatic force**.
6. The force exerted by a magnet is called **magnetic force**.
   (i) Magnetic cranes are used to lift heavy iron objects (attraction).
   (ii) Magnetic force lifts a Maglev train slightly above the track to reduce noise and friction (repulsion).
7. The pull of the earth is called **gravitational force**. It is due to the gravitational force that an object falls to the ground, when released from a height.
8. **Determining net force.** The net force is the combination of all the forces acting on an object:
   (i) When forces act in the same direction, the net force is the sum of all the forces and acts in the same direction as the individual forces.
   (ii) When two forces act in opposite directions, the net force is obtained by subtracting the smaller force from the larger force and acts in the direction of the larger force.

**9.** (i) When the forces acting on a body nullify each other and produce a net force of 0N, the forces are said to be balanced.

(ii) When the net force on an object is not 0N, the forces on the object are ***unbalanced.*** Unbalanced forces produce a change in motion, such as a change in speed or a change in direction.

**10.** The **weight of a body** is the gravitational pull of the earth on the body, *i.e.*, the weight of a body is the force with which the earth attracts it towards it.

**On earth a 100 g mass has a weight of about 1 N.** Thus, the force required to hold up a 100 g orange or apple in hand is 1 N.

**11.** If the mass of a body is ***m*** and acceleration due to gravity is ***g***, then the weight ***W*** of the body is given by the formula

$$W = m \times g$$

*i.e.*, **weight = mass × acceleration due to gravity**

**12.** The acceleration due to gravity is $g$ = **980 cm/sec², *i.e.*, 9.8 m/sec²**. Hence, we have

**Weight of a mass of 1 kg** = 1 kg × 9.8 = **9.8 N = 10 Newtons (approx.)**

Since weight is the force exerted on a mass, it can also be expressed in kilogram force (kgf) also called kilogram weight (kg wt).

$$\mathbf{1\ kgf = 9.8\ N = 10\ N\ (approx.)}$$

Note that mass gives the measure of how much matter a body contains whereas weight gives the measure of the gravitational pull on the mass of the body.

**13.** Since the weight of a body of given mass depends on the gravitational pull, it will change slightly as we move from one place to another on the earth because gravity changes from place to place. Thus, **our weight is maximum at the poles and minimum at the equator.**

Also, since the gravitational pull of moon is about one-sixth of that of the earth, the weight of a body of 1 kg mass on moon would be roughly $\frac{1}{6}$th of 9.8 N = **1.6 N.**

That is why a person can jump six times higher on the moon than on the earth.

Similarly, it will vary on other planets. The weight of a body on Jupiter would be 2.364 times that on the earth. Thus, a person having 50 kg weight on earth will weigh 118.2 kg wt. on Jupiter.

In space, the acceleration due to gravity is zero, and so, the weight of the body in space shall be zero. ($w = mg = m \times 0 = 0$).

**14. Weightlessness.** It is the state when a person does not feel any sensation of weight. It happens when there is no contact with any surface to oppose gravity. Astronauts appear to be weightless while they are floating inside the space shuttle but they are not weightless. It is impossible for any object to be weightless anywhere in the universe. Why?*

**15. Pressure.** A force can produce a high or low pressure depending on the area that the force exerts. When a pin is pushed into a board, the tip of the pin exerts a high pressure on the board because the force applied is concentrated on a small area of the pin. The area of objects must be reduced to increase pressure.

Low pressure is useful when walking or moving on soft snow and soft ground. The area of objects must be increased to reduce pressure.

---

* Suppose you travelled in space far away from all the stars and planets. The gravitational force acting on you would be very small because the distance between you and other objects would be very large. But you and all the other objects in the universe would still have mass. Therefore, gravity would attract you to other objects–even if just slightly so you would still have weight.

Trucks and buses have broader tyres as compared to cars and bicycles as larger area reduces the pressure exerted on the ground. Water tanks are broader at the bottom than at the top to reduce pressure. **Pressure is defined as the force acting on a unit area**

$$\textbf{Pressure} = \frac{\textbf{Force}}{\textbf{Area}}$$

The SI unit for pressure is **pascal (Pa)** or **N/m²**, where 1 Pa = 1 N/m², *i.e.*, **one pascal is the force of one newton exerted over an area of one square metre.**

Note that Pressure $= \frac{\text{Force}}{\text{Area}} \Rightarrow$ ***Force = Pressure × Area.***

The relation Pressure $= \frac{\text{Force}}{\text{Area}}$ shows that the ***pressure is inversely proportional to the area***. If area is less then pressure will be greater.

**Ex. 1.** ***Find the pressure exerted by an object that has an area of 0.05 m² and a weight of 10 N ?***

**Sol.** Pressure $= \frac{10\,\text{N}}{0.05\,\text{m}^2} = \frac{10\times100}{5}$ N/m² = **200 Pa.**

**Ex. 2.** ***Find the weight of a rock that has an area of 20 m² and which exerts a pressure of 500 Pa.***

**Sol.** Weight (force) = 500 Pa × 20 m² = **10000 N.**

**16. Atmospheric pressure.** The pressure exerted by the air on the surface of the earth is called the atmospheric pressure. This is given by

$$p = hdg$$

where $h$ is the height of atmosphere above earth's surface, $g$ is the acceleration due to gravity and $d$ is the density of air which is assumed to be uniform.

**17. Fluid pressure** at depth $h$ is given by $p = hdg$ where $d$ is the density of the liquid and $g$ is the acceleration due to gravity. The formula shows that

(i) ***Liquid pressure depends upon the height of the liquid column and its density. It does not depend upon area of base. It increases with depth.***

(ii) Liquid exerts lateral pressure.

(iii) Liquid pressure acts uniformly on any area.

**18. Buoyant force.** The upward force experienced by a body when partly or wholly immersed in a fluid is called **upthrust** or **buoyant force**, whereas the phenomenon responsible for this force is called **buoyancy**. ***The buoyant force or upthrust depends on the volume of the body and the density of the liquid.*** It increases with the increase in volume of the body immersed in fluid, and hence with the increase in the volume of the fluid displaced by the body. It also increases with the increase in density of the fluid.

**19.** When a body is immersed in a liquid, two forces act on it.

(i) weight $W_1$, of the body acting downward

(ii) buoyant force (upthrust) $F$.

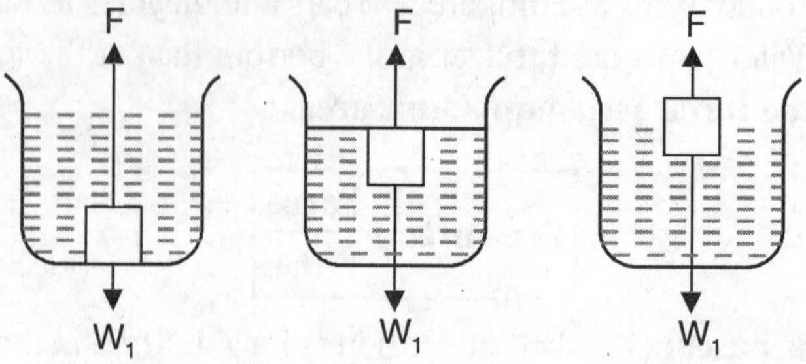

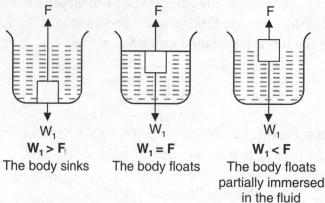

**$W_1 > F$** The body sinks
**$W_1 = F$** The body floats
**$W_1 < F$** The body floats partially immersed in the fluid

The body will sink if $W_1 > F$, float if $W_1 = F$ and float partly immersed if $W_1 < F$.

Thus, **the apparent weight of a body fully or partially immersed will always be less than the actual weight**. This is the reason why it is easier to lift a heavy stone under water than outside. When you draw out water in a bucket from a well, you do not feel its weight so long as it is inside water. The moment it is taken out of water, you feel its weight and have to pull it up by applying force.

**20. Archimedes principle.** It states that "***when a body is wholly or partially immersed in a liquid, there is apparent loss in the weight of the body, which is equal to the weight of the displaced liquid by the body.***"

**21. Law of floatation.** It states that an object floats in a liquid if the weight of the object is equal to the weight of the liquid displaced by the submerged part of the body. In such a case, the apparent loss in the weight of the object is zero. Thus, a piece of wood floats in water because the weight of water displaced by it is equal to its own weight.

**22. Pascal's principle.** ***It states that an enclosed fluid exerts pressure equally in all directions***. Thus, if the water - pumping station in the town increases the water pressure by 30 Pa, it will be increased by the same amount at all locations, whether 50 metres away or 2 km away.

## Question Bank-2(a)

**1.** A force may cause in an object
(a) change in inertia (b) change in mass
(c) change in weight (d) none of these

**2.** A force
(a) is expressed in Newtons.
(b) can cause an object to speed up, slow down, or change direction.
(c) is a push or a pull
(d) all of the above.

**3.** Identify the following forces :
(a) The force used to light up a match.
(b) The force which changes the direction of a compass needle.
(c) The force which pulls a meteorite in space towards a moon.

**4.** Identify the forces that are operating while you are riding a bicycle.

**5.** The unit of force is
(a) kg (b) N
(c) Pa (d) $N/m^2$

**6.** Mass is measured in
(a) litres (b) Newtons
(c) centimetres (d) kilograms

**7.** Your pair of shoes has a mass of 780 g. If each shoe has exactly the same mass, what is the weight of each shoe?

[**Hint.** 100g of mass has a weight of 1N on earth]

**8.** Friction is a/an
(a) self-adjusting force

(b) necessary evil
(c) important force in daily life
(d) all of the above

9. A nugget of gold is placed in a graduated cylinder that contains 100 mL of water. The water level rises to 350 mL after the nugget is added to the cylinder. What is the volume of the gold nugget?

10. Which of the following statements about weight is true?
(a) Weight is a measure of the gravitational force on an object.
(b) Weight varies depending on where the object is located in relation to the Earth.
(c) Weight is measured by using a spring scale.
(d) All of the above.

11. When a bicycle travels on a rough surface, its speed
(a) increases (b) decreases
(c) remains the same (d) none of these

12. It is difficult to walk on ice because of
(a) little friction (b) absence of inertia
(c) more inertia (d) more friction

13. The S.I. unit of pressure is
(a) Newton/metre$^2$ (b) Newton
(c) metre$^2$ (d) kg

14. Which of the following is not an example of upthrust?
(a) Upward force in water
(b) Force acting on a stone
(c) Upward force in milk
(d) Upward force in oil

15. The weight of a body is equal to
(a) mass × gravity (b) mass/gravity
(c) gravity/mass (d) none of these

16. Which of the following may happen when an object receives unbalanced forces?
(a) The object changes direction.
(b) The object changes speed.
(c) The object starts to move.
(d) All of the above.

17. A boy pulls a wagon with a force of 6 N east as another boy pushes it with a force of 3 N east. What is the net force?
(a) 2 N (b) 18 N
(c) 9 N (d) 3 N

18. Biological forces, mechanical forces and frictional forces are the examples of :
(a) contact forces
(b) non-contact forces
(c) both (a) and (b)
(d) neither (a) nor (b)
(e) none of these

19. What must you know in order to calculate the gravitational force between two objects?
(a) their mass
(b) the distance between them
(c) both of the above
(d) none of these

20. If you are in a spacecraft that has been launched into space, your weight would
(a) increase because gravitational force is increasing.
(b) increase because gravitational force is decreasing.
(c) decrease because gravitational force is decreasing.
(d) decrease because gravitational force is increasing.

21. The gravitational force between 1 kg of lead and Earth is ________ the gravitational force between 1 kg of melon and earth.
(a) greater than
(b) the same as
(c) less than
(d) none of the above

22. You may sometimes hear on the radio or on TV that astronauts are weightless in space. Explain why this statement is not true.

23. An iron ball and a wooden ball of the same radius are released from a height $H$ in vacuum. The times taken by both of them to reach the ground are
(a) roughly equal
(b) unequal

(c) exactly equal
(d) in the inverse ratio of their diameters

**24.** Where would you weigh the most?
(a) on a boat (b) on the space shuttle
(c) on the moon (d) on Venus

**25.** If Earth's mass is doubled without changing its size, your weight would
(a) increase because gravitational force increases
(b) decrease because gravitational force increases
(c) increase because gravitational force decreases
(d) not change because you are still on Earth.

**26.** Your friend thinks that there is no gravity in space. How could you explain to your friend that there must be gravity in space.

**27.** The gravitational force on Jupiter is approximately 2.3 times the gravitational force on Earth. If an object has a mass of 80 kg and a weight of 600 N on Earth, what would the object's mass and weight on Jupiter be?

**28.** We can hold a pen due to the
(a) force of gravity
(b) force of friction
(c) force of weight
(d) work done by our muscles

**29.** The weight of a body on the surface of the earth is 10 kg. Its weight at the centre of the earth is
(a) 0 kg (b) 5 kg
(c) 10 kg (d) Infinite

**30.** Explain why it is your weight and not your mass that would change if you landed on Mars.

**31.** Flat-footed camels can walk easily in sandy deserts because
(a) pressure on the sand is decreased by increasing the area of the surface in contact.
(b) pressure on the sand is increased by increasing the area of the surface in contact.
(c) pressure on the sand is decreased by decreasing the area of the surface in contact.
(d) pressure on the sand is increased by decreasing the area of the surface in contact.

**32.** Why is one end of a sewing needle made pointed?
(a) to increase the force
(b) to increase the pressure
(c) to decrease the weight
(d) none of the above.

**33. Give reasons for the following:**
(1) Magicians are able to lie down on a bed of nails.
(2) Walls of the water reservoir of a dam have to be made wider at the bottom.
(3) The tips of the cutting tools are made wedge-shaped.
(4) The feet of an elephant are large.
(5) It is easier to move with skis on snow.
(6) A man walking on the street slips on a banana skin.
(7) The soles of our shoes wear out gradually.
(8) It is difficult to walk on well polished floor or ice.

**34.** *A* and *B* are two objects with masses 10 kg and 40 kg respectively. Then
(a) *A* has more inertia than *B*
(b) *B* has more inertia than *A*
(c) *A* and *B* have the same inertia
(d) Neither *A* nor *B* has any inertia

**35.** A steel chain consists of 50 links connected to each other. The chain is pulled by applying force *F* at each of its ends. What is the magnitude of force induced in each link?
(a) $F/50$ (b) $F$
(c) $50\ F$ (d) $F/25$

**36.** A long chain having 100 identical steel links is to be pulled from its two ends by a force of 3000 Newtons. The chain is elongated by adding another 50 links to it. Which one of the following is the correct statement?
(a) The force induced in each link will be equal to 20 Newtons.

(b) The force induced in each link will be equal to 30 newtons.

(c) The force induced in each link will be equal to 3000 newtons.

(d) The force induced in each link cannot be determined unless its dimensions are given.

**37.** Pressure cannot be measured in

(a) $N/m^2$ (b) bar

(c) Pa (d) kg wt

**38.** When the force remains constant and area is less, the pressure will

(a) increase

(b) decrease

(c) first increase then decrease

(d) first decrease then increase

**39.** When depth increases the pressure

(a) increases

(b) remains constant

(c) first decreases then increases

(d) decreases

**40.** When a constant force is applied to a body, it moves with uniform

(a) speed (b) velocity

(c) acceleration (d) momentum

**41.** A piece of wood is floating in water. If we heat the water, the piece of wood will

(a) increase in size

(b) rise a little

(c) float at the same level

(d) sink a little

**42.** The bottom of a dam is made thick as:

(a) the water exerts low pressure on bottom wall

(b) it is a custom

(c) it looks beautiful

(d) the water exerts more pressure on bottom of the wall

**43.** The buoyant force depends on the

(a) depth of a liquid

(b) colour of a liquid

(c) density of a liquid

(d) none of these

**44.** A body of wood has a weight $W$ and volume $V$. The apparent weight of this body after making it float on water will be

(a) $\frac{W}{V}$ (b) $W \times V$

(c) $W$ (d) Zero

**45.** Where would the pressure of sea water be maximum?

(a) Below 100 metres from the surface.

(b) Below 80 metres from the surface.

(c) Below 105 metres from the surface.

(d) Below 10 metres from the surface.

**46.** If the force remains constant and the area is doubled, then the pressure will be

(a) half (b) three times

(c) two times (d) none of these

**47.** Two objects losing the same weight when immersed in water must have the same

(a) weight in water (b) volume

(c) weight in air (d) density

**48.** The hot air balloon rises because it is

(a) denser than air

(b) less dense than air

(c) equally dense as air

(d) The given statement is wrong

**49.** An ice cube is floating in a glass of water. How will the water level in the glass be affected when the ice cube melts?

(a) It will rise

(b) It will go down

(c) It will remain unchanged

(d) It will first go up but later on it will go down.

**50.** The balls of iron and aluminium of same diameter are dipped in water. Which of the following is the correct statement?

(a) The upthrust on iron ball will be more than on the aluminium ball.

(b) The upthrust on aluminium ball will be more than on the iron ball.

(c) The upthrust on both will be the same.

(d) None of the above.

**51.** The height of mercury which exerts the same pressure as 20 cm of water column, is

(a) 1.47 cm (b) 14.8 cm

(c) 148 cm (d) none of these

[**Hint.** Pressure = $hdg$

$\therefore$ Let height of mercury be $h$ cm.

Then $h \times 13.6 \times g = 20 \times 1 \times g$

$\Rightarrow h = \dfrac{20}{13.6} = 1.47$ cm]

**52.** An elephant weighing 60,000 N stands on one foot covering an area of 1000 $cm^2$. Find the pressure exerted by the elephant on the ground.

**53.** Find the pressure exerted by a girl weighing 500 N standing on one stiletto heel of area 1$cm^2$.

**54. Create a concept map using the following terms.**

*contact forces, non-contact forces, friction, push, pull, upthrust, twist, gravitational force, magnetic force, electrostatic force, atmospheric pressure, fluid pressure, pascal.*

**55. Solve the following crossword using the given clues :**

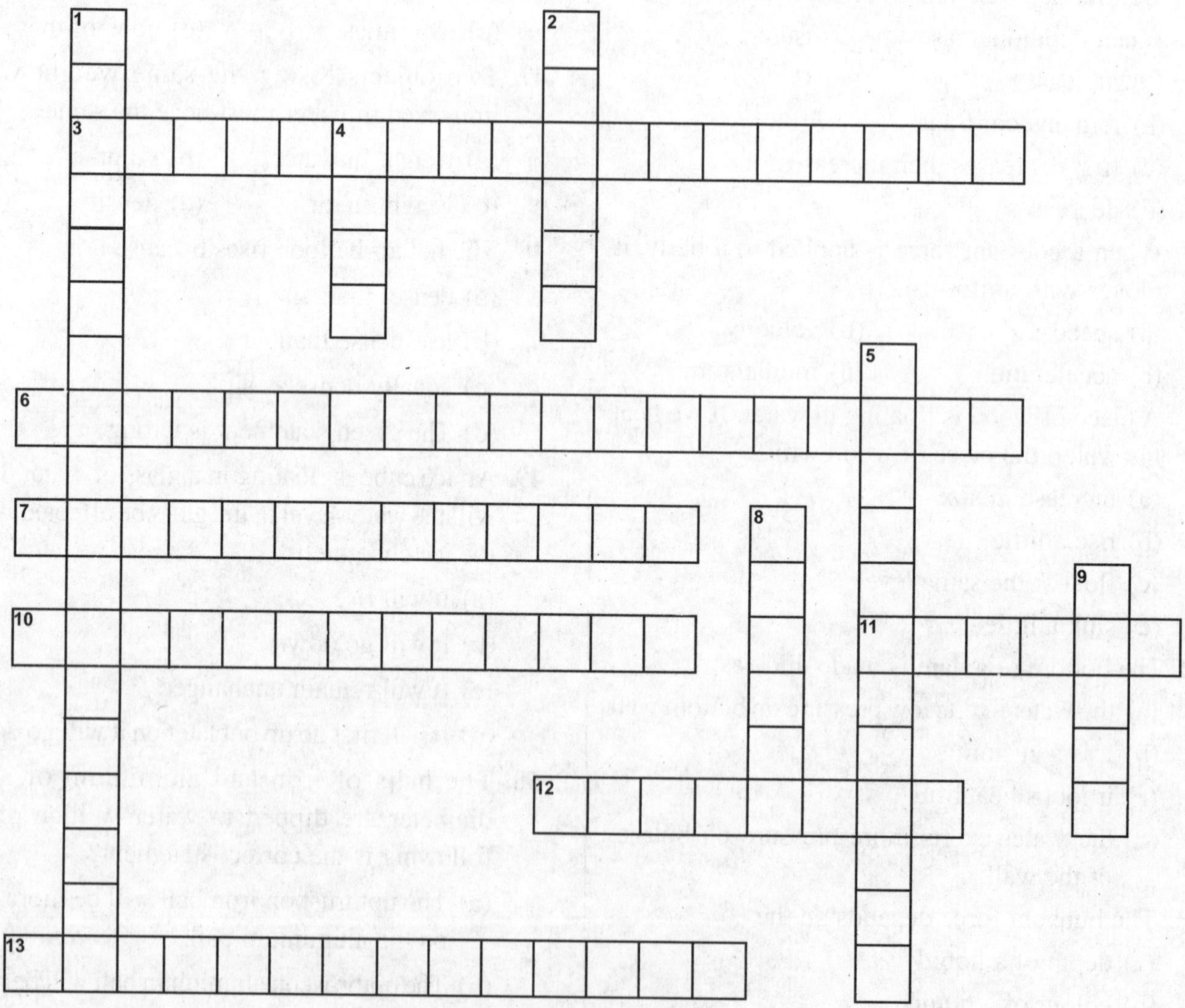

### ACROSS

3. Principle relating to the apparent weight of a body when immersed in water.
6. Pressure caused by the weight of the air
7. Forces that require physical contact between objects
10. The force that exists between two magnets.
11. Unit of force
12. The resistance to movement that occurs when two bodies are in contact
13. The state when a person does not feel any sensation of weight.

### DOWN

1. The pull of the earth
2. SI unit of pressure
4. Measure of the amount of matter in a physical body
5. The upward force that a fluid exerts on an object that is immersed in the fluid.
8. The force with which the earth pulls a body towards its centre.
9. Something that tends to cause movement of a body.

## Answers

**1.** (c) **2.** (d)

**3.** (a) Frictional force (b) Magnetic force (c) Gravitational force

**4.** Feet exert force on the pedals; the tyres exert force on the ground; fingers exert force on the hand brakes; the brake pads exert force on the wheel rims.

**5.** (b) **6.** (d) **7.** 3.90 N **8.** (d)

**9.** 250 $cm^3$ (1 mL = 1 $cm^3$) **10.** (d) **11.** (b)

**12.** (a) **13.** (a) **14.** (b) **15.** (a)

**16.** (d) **17.** (c) **18.** (a) **19.** (c)

**20.** (c) **21.** (b)

**22.** An astronaut weighs less in space than on Earth because of the astronaut's increased distance from Earth. But an astronaut is not weightless because there are still gravitational forces between the astronaut and all other objects in the universe.

**23.** (c) **24.** (a) **25.** (a)

**26. Sample answer:** You can tell your friend that there must be gravity in space because gravity holds the planets in orbit around the sun.

**27.** mass = 80 kg, weight = 1380 N

**28.** (b) **29.** (a)

**30.** Your weight would change if you landed on Mars because the gravitational force on Mars is different from the gravitational force on Earth. But your mass would not change because the amount of matter in your body would not change.

**31.** (a) **32.** (b)

**33.** (1) The magician's weight is fairly large, but because there are hundreds or even thousands of nails, the pressure (the amount of force exerted on a given area) from each nail is not enough to cause harm to the magician's skin.

(2) Because pressure due to liquid increases with depth.

(3) To minimise the area of the cutting surface to be used for cutting. This helps in exerting maximum pressure with a minimum force applied by the cutting tool.

(4) The weight of the elephant is quite large and to balance it the area of the feet is kept more.

(5) The surface area of a ski is almost 20 times as large as that of the sole of shoes. Hence, on skis, a person is able to exert pressure on each square centimetre of the surface area of the snow which is almost 20 times as small as that when he walks with shoes on.

(6) The banana skin reduces the friction between the foot and the floor.

(7) Due to friction between the soles of shoes and the road.

(8) The friction on polished surfaces or ice is very small and is not enough to prevent us from slipping.

**34.** (b) **35.** (b) **36.** (a) **37.** (d)
**38.** (a) **39.** (a) **40.** (c) **41.** (d)
**42.** (d) **43.** (c) **44.** (d) **45.** (c)
**46.** (a) **47.** (b) **48.** (b) **49.** (c)
**50.** (c) **51.** (a)

**52.** $\text{Area} = 1000\ \text{cm}^2 = \dfrac{1000}{10000}\ \text{m}^2 = \dfrac{1}{10}\ \text{m}^2$

$$\therefore \text{Pressure} = \frac{\text{Force}}{\text{Area}} = \frac{60000\ \text{N}}{\frac{1}{10}\ \text{m}^2}$$

$$= 6{,}00{,}000\ \text{N/m}^2$$

**53.** $\text{Area} = 1\ \text{cm}^2 = \dfrac{1}{10{,}000}\ \text{m}^2$

$$\therefore \text{Pressure} = \frac{500\ \text{N}}{\frac{1}{10{,}000}\ \text{m}^2} = 50{,}00{,}000\ \text{N/m}^2$$

**54. Concept map**

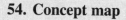

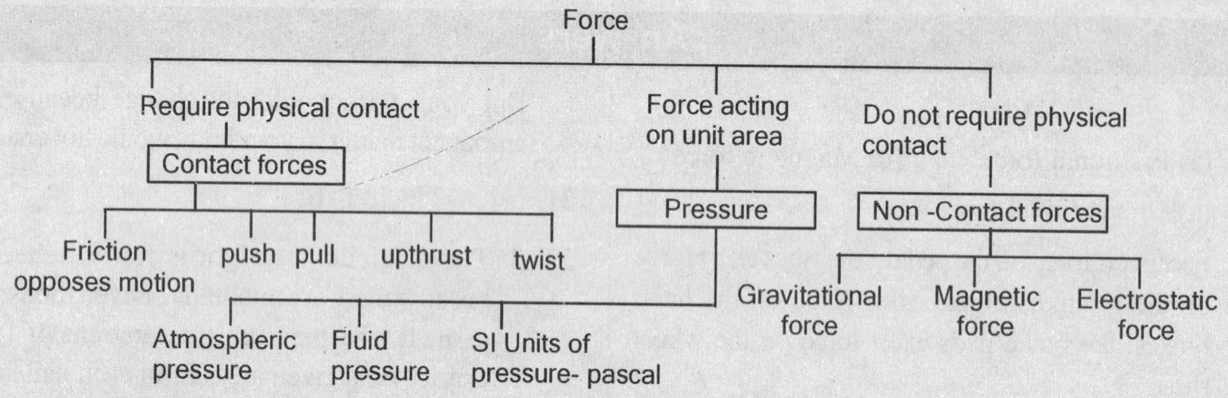


**55.**

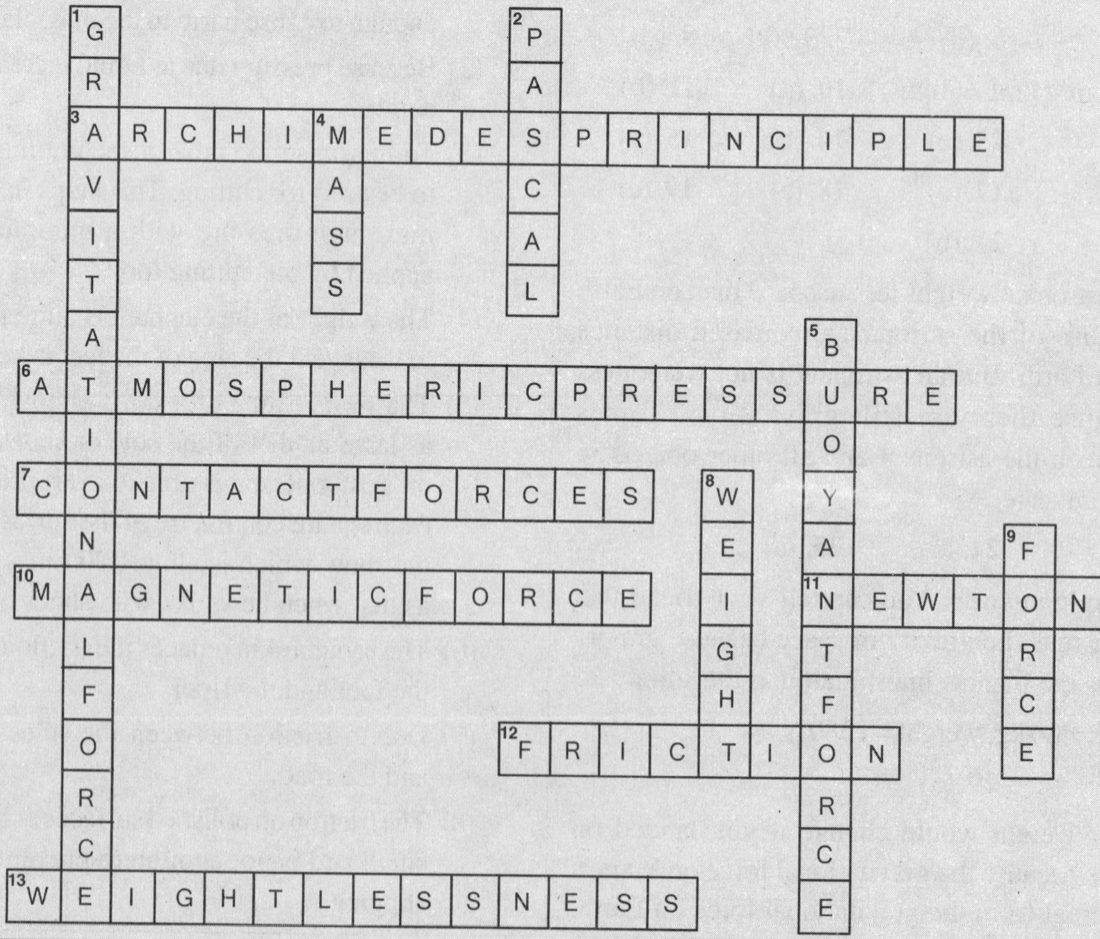

## Self Assessment Sheet-2

1. **Give reasons for the following.**
   (i) Spikes are provided in the shoes of players.
   (ii) Grooves are provided in the shoes of players.
2. 1 N force is exerted on the palm of a child and the same force is also exerted on the palm of a man. Who of the following will feel more pressure ?
   (a) Man
   (b) Child
   (c) Both the man and the child
   (d) None of the above
3. One boy is carrying a bucket of water by his one hand and wooden cube in his other hand. If he places the wooden cube in the bucket of water, then he will feel
   (a) less weight (b) more weight
   (c) same weight (d) none of these
4. What force is required to exert a pressure of 40,000 Pa on an area of 1 $cm^2$ ?
5. Find out the area of a body which experiences a pressure of 60,000 Pa by a force of 500 N ?
6. Why is it easy to slip when there is water on the floor?
   (a) The water is a lubricant and reduces the friction between your feet and the floor.
   (b) The friction between your feet and the floor changes from kinetic to static friction.
   (c) The water increases the friction between your feet and the floor.
   (d) The friction between your feet and the floor changes from sliding kinetic friction to rolling kinetic friction.
7. If a rock is brought from the surface of the moon
   (a) its mass will change
   (b) its weight will change, but not mass
   (c) both mass and weight will change
   (d) its mass and weight will remain the same
8. The pressure on earth will be least when the man is
   (a) lying
   (b) sitting
   (c) standing on one foot
   (d) standing on two feet
9. The force which slows down or stops the ball kicked by you is
   (a) gravitational force (b) frictional force
   (c) muscular force (d) mechanical force
   (e) none of these
10. The force required to lift 200 g of mass vertically against the force of gravity is expressed in
    (a) Newton (b) gram force
    (c) kilogram force (d) all of these
    (e) none of these

## Answers

1. (i) Since the players and athletes have to run fast, therefore, greater friction is required between the soles of their shoes and ground to prevent slipping. This is achieved by providing spikes in the soles of shoes. Being pointed nails, spikes get into the ground and increase the friction between the shoe and the ground.
   (ii) The grooves in the tyres increase the friction between tyre and road and hence prevent the skidding (or slipping) of the moving vehicle on a slippery road.

**2.** (b) **3.** (b) **4.** 4 N **5.** 120 $m^2$
**6.** (a) **7.** (b) **8.** (a) **9.** (b)
**10.** (a)

# Chapter 3

# WORK AND ENERGY

## KEY FACTS

**1. Work is done when a force causes an object to move.**

Thus, two things happen when work is done on an object

(1) the object moves as a force is applied, and

(2) the direction of the object's motion is the same as the direction of the force.

**Illustrations : 1.** No work is done if you stand waiting for someone for, say, half an hour. Since there is no movement, so no work is done.

**2.** Work is done on a suitcase when you lift it off the ground. If you run with a suitcase in your hand, you are not doing work on the suitcase because you are applying a force to hold the suitcase up, but the suitcase is moving forward. For work to be done on an object, the object must move in the *same direction* as the force.

**3.** The amount of work ($W$) done in moving an object is calculated by multiplying the force applied on the object and the distance moved through by the object. Thus,

**Work done = Force applied × distance moved by the object in the direction of the force**

*i.e.,* $\boxed{W = f \times d}$

The S.I. unit of work is **joule** denoted by **J**.

> **1 Joule** ***is equivalent to the amount of work done by a force of 1 N acting on a body that moves through a distance of 1 m in the direction of the force.***

1 joule is a small unit and usually we express work done in **kilojoules (kJ)**.

**1 kJ = 1000 J.**

**Ex. 1.** ***Using a force of 20 N, a person pushes a shopping cart 8 m. How much work does the person do?***

**Sol.** Work done by the person $= f \times d = 20 \text{ N} \times 8 \text{ m}$

$= \mathbf{160 \text{ J}}$.

**Ex. 2.** ***(Conceptual problem) A chair weighing 60 N is lifted to a height of 0.8 m and then carried 6 m across the room. How much work is done on the chair?***

**Sol.** Work done $= 60 \text{ N} \times 0.8 \text{ m} = 60 \text{ N} \times \frac{8}{10} \text{ m}$

$= \mathbf{48 \text{ J}}$.

No further work is done on the chair once it has been lifted, because the direction in which the person walks is perpendicular to the direction in which the chair is lifted.

2. **Energy is the capacity or ability to do work**. After playing for a long time or running a long distance, you get tired because you have consumed some of the energy in doing the work and have less energy now. After waking up in the morning you feel energetic. This means you have more capacity to do work and can run faster or over a longer distance.

***The greater the energy in a body, the greater is the work it is capable of doing.***

3. **Energy can be transferred from one object to another object**. Imagine two players playing a game of tennis. The tennis player tosses the ball into the air and then slams it with his racket. The ball flies toward his opponent, who swings his racket at the ball and the ball goes into the net, causing it to shake.
The tennis player does work on his racket by exerting a force on it, the racket does work on the ball, and the ball does work on the net. When one object does work on another, energy is transferred from the first object to the second object. This energy allows the second object to do work. So, work is a transfer of energy. A moving bat can set a ball in motion, because energy gets transferred from it to the ball.

4. **Unit of energy.** The unit of energy is also joule, the same as that of work.

5. **Kinetic energy.** ***The energy of an object that is due to the object's motion is called kinetic energy.*** You have seen that in tennis, energy is transferred from the racket to the ball. As it flies over the net, the ball has kinetic energy. All moving objects have kinetic energy. Like all forms of energy, kinetic energy can also be used to do work.
Kinetic energy depends on two quantities.
(i) speed or velocity
(ii) mass
Faster a body moves, the more kinetic energy it has. Also, more the mass, the higher the kinetic energy is. Kinetic energy of an object of mass $m$ moving with speed $v$ is given by the formula

$$\textbf{Kinetic energy} = \frac{1}{2} \times \textbf{mass} \times (\textbf{speed})^2$$

*i.e.,*

$$\textbf{K.E.} = \frac{1}{2} mv^2$$

**Ex.** ***What is the kinetic energy of a vehicle having a mass of 2000 kg and moving at 20 m/s.***

**Sol.** K.E. of the vehicle $= \frac{1}{2} mv^2$

$$= \frac{1}{2} \times 2000 \text{ kg} \times (20 \text{ m/s})^2$$

$$= \frac{1}{2} \times 2000 \text{ kg} \times 400 \text{ m}^2/\text{s}^2 = \textbf{4,00,000 J.}$$

6. **Potential energy**. ***Potential energy is the energy an object has because of its position.*** It is the stored energy or the energy stored in those objects which are not moving. It is the energy which is inactive at the moment but has the potential to do work. As we bend a bow, raise a hammer, pull the pendulum to a side, stretch rubber in a catapult, we do work in all these cases and this work is turned into potential energy.

7. **Gravitational potential energy.** When an object is lifted, work is done on it by applying a force against the force of gravity. In doing so, energy is transferred to the object as gravitational potential energy. The gravitational potential energy of a body depends on :

(i) the mass of the body,

(ii) the height through which it is raised from the ground, and

(iii) the gravity.

The greater the mass of the body, or the higher it is from the ground or more the gravity, the more would be the gravitational potential energy.

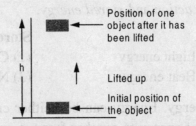

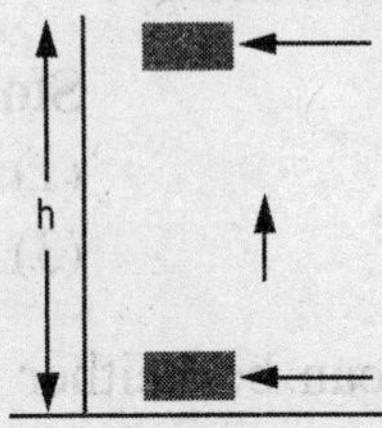


Let an object of mass $m$ be lifted through a height $h$.

Then, work done in raising the mass through height $h$ = force × displacement

= weight × height

= $mg \times h$

Hence, by definition of potential energy,

G.P.E. = work done in raising mass $m$.

*i.e.*, **G.P.E. = weight × height = *mgh***

Because weight is expressed in newtons and height in metres, gravitational potential energy is expressed in newton-metres, or joules (J).

We may also think of gravitational potential energy as being equal to the work that would be done by the object if it were dropped from height $h$.

**Ex.** ***An object of mass 8 kg is placed at a height of 5 m above the ground.***

***(a) What kind of energy does this object possess ?***

***(b) Find the amount of energy possessed by the object. (Take g = 9.8 m/s²)***

**Sol.** (a) Gravitational potential energy.

(b) G.P.E. = $mgh$

= 8 kg × 9.8 m/s$^2$ × 5 m = **392 J**.

8. **Elastic potential energy** is the energy stored when objects are stretched, compressed or bent like squeezing of a ball, winding up of a watch. Elastic potential energy can be stored by stretching a rubber band. It is released when the rubber band is let go and becomes kinetic energy.

9. **Mechanical energy.** Mechanical energy is the total energy of motion and position of an object. Both potential energy and kinetic energy are kinds of mechanical energy. Mechanical energy can be all potential energy, all kinetic energy, or some of each. The following equation can be used to calculate mechanical energy.

**Mechanical energy = Potential energy + Kinetic energy**

The mechanical energy of an object remains the same unless some of its energy is transferred to another object. But the kinetic or potential energy may increase or decrease.

**Illustration.** As a juggler moves a ball with his hand, he does work on the ball to give it kinetic energy. As soon as the ball leaves his hand, the ball slows down as it moves upward and the kinetic energy starts changing into potential energy. Eventually, all the ball's kinetic energy changes into potential energy, and it stops moving upward. Now, as it starts falling downward, it starts picking up speed and its potential energy starts changing into kinetic energy.

**10. Other forms of energy.** Besides mechanical energy, energy can come in a number of other forms. These can be classified as *energy in action* and *stored energy*.

**Energy in action**

(1) Sound energy (2) Light energy
(3) Electrical energy (4) Heat energy

**Stored energy**

(1) Chemical energy (2) Magnetic energy
(3) Nuclear energy

**11. Law of conservation of energy**. Energy can be neither created nor destroyed. However, it can be transformed from one form of energy to another form of energy and whenever this transformation of energy of one form to the other takes place, the total energy always remains the same or it remains conserved.

**Illustration** : The figure given below shows a simple pendulum suspended from a rigid support *O*. *A* is its resting position and when it is displaced to one side and then released, it swings from one side to the other, covering equal distance and reaching equal height on either side of its mean position *A*.

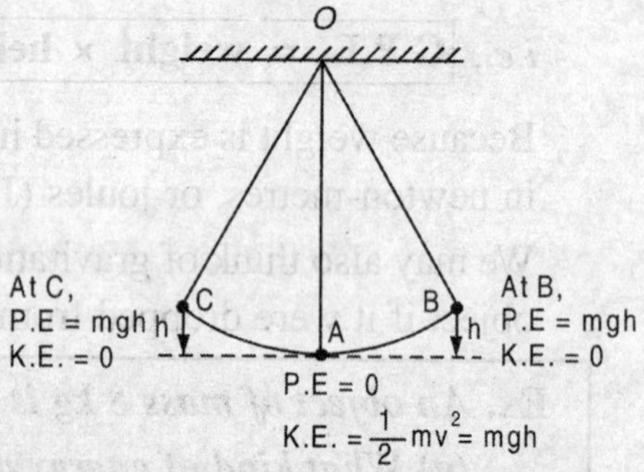


Neglecting the frictional force between the bob and the surrounding air, we have

At *B* or *C*, max. P.E. = $mgh$ and K.E. = 0

As the bob swings back from *B* or *C* to *A*, the P.E. decreases and the kinetic energy increases. At *A*, its total mechanical energy is in the form of kinetic energy and the potential energy is zero.

At an intermediate position like between *A* and *B* or *A* and *C*, it has both the kinetic energy as well as the potential energy whose sum remains the same, *i.e.*, $mgh$.

The conservation law is strictly true in vacuum where there is no friction due to air.

**12. Energy conversions.** An energy conversion is a change from one form of energy to another. Any form of energy can change into any other form of energy. Often, one form of energy changes into more than one form.

**Examples :**

(i) Table fan → Electrical to mechanical
(ii) Microphone → Sound to electrical
(iii) Wind mill → Mechanical to electrical
(iv) Burning of a candle → Chemical to light

**13. Sources of energy.** Energy is obtained from various sources like from the sun, from flowing water, from winds, from biomass, fossil fuels, sea waves etc.,

**14. Renewable sources of energy.** The sources of energy such as sun, wind, water, etc., which never run out as they are used or are naturally replaced more quickly than they are used are called renewable sources.

**15. Non-renewable sources of energy.** The sources of energy which cannot be replaced once they are used up are called non-renewable sources of energy, *e.g.,* petroleum and coal.

**16. Fossil fuels.** Oil and natural gas, as well as coal, are the most common fossil fuels. These are energy resources that formed from the burried remains of plants and animals that lived millions of years ago. These plants stored energy from the sun by photosynthesis. Animals used and stored this energy by eating the plants. So fossil fuels are concentrated forms of the sun's energy.

**17. Biomass.** Plants use and store energy form the sun. Organic matter, such as plants, wood, and waste, that can be burnt to release energy is called biomass.

## Question Bank–3

**Conceptual Questions :**

**1. Fill in the blanks:**

> **Word Bank**
>
> *Work, energy, potential energy, wind, electrical, light transformed, conservation, position, newton-metre, joule, kinetic energy, joules, renewable mechanical energy, gravitational potential energy, non-renewable mechanical, chemical.*

(a) _____ is said to be done when a force causes an object to move in the direction of the force.

(b) The S.I. unit of work is _____ which is more simply called _____.

(c) 1 _____ is the amount of work done by a force of 1N acting through a distance of 1 _____ in the direction of the force.

(d) _____ is the ability to do work and is expressed in units of _____.

(e) _____ is the energy which a body possess by virtue of its position, shape or condition.

(f) _____ is the energy which a body possess because of its motion.

(g) _____ is the sum of kinetic energy and potential energy.

(h) The energy stored in an object when it is lifted above the surface of earth is called _____.

(i) Energy can be _____ from one form to another.

(j) A microphone converts sound energy to _____ energy.

(k) The law of _____ of energy states that energy can neither be created nor destroyed.

(l) Burning of wood converts _____ energy to heat and light energy.

(m) Washing machine converts electrical energy to _____ energy.

(n) There is practically limitless supply of _____ source of energy.

(o) Plants convert _____ energy into chemical energy.

(p) Energy and work are expressed in units of _____.

(q) Wind turbines convert _____ energy into electrical energy.

(r) _____ sources of energy are more quickly used up than they are being replaced.

**2. Name the type of energy (K.E. or P.E.) possessed in the following cases :**

(a) A moving cricket ball

(b) A compressed spring

(c) A moving bus

(d) The bob of a simple pendulum in its extreme position

(e) A stretched bow

(f) Water stored at a height

(g) The bob of a simple pendulum at its mean position.

(h) A stone placed at the roof

(i) Flowing water

(j) A suitcase kept on the head.

(k) A wound up watch spring.

**3. Answer true or false.**

(a) Work is done when we try to push a wall.

(b) Falling water can be used to produce electrical energy.

(c) Photosynthesis converts solar energy to chemical energy.

(d) A cell or battery converts electric energy into chemical energy.

(e) K.E. of an object of mass $m$ moving with velocity $v$ is equal to $\frac{1}{2}mv$.

(f) The S.I. unit of work is Newton.

(g) Energy exists in different forms.

(h) Energy can be converted from one form to another.

(i) P.E. of an object placed at a height of $h_1$ is more than an object placed at height $h_2$, if $h_1 > h_2$.

(j) Sun is the ultimate source of energy.

(k) Work is done by a coolie standing still with heavy luggage on his head.

**4.** How do the meanings of the two terms '*Work* and *Joule*' differ?

**5.** What are the two things that must happen for work to be done?

**6.** (a) Work is done on a ball when a player kicks it. Is the player still doing work on the ball as it flies through the air? Explain.

(b) You lift a chair that weighs 30 N to a height of 0.8 m and carry it 20 m across the hall. How much work is done by you on the chair ?

**7.** From the ground floor a man comes up to the fourth floor of a building using a staircase. If the man comes up to the same floor using an elevator, neglecting friction, compare the work done by the man in the two cases.

**8. Answer the following questions giving explanation for your answers :**

(a) How is work done related to applied force?

(b) When an arrow is shot from a bow, it has kinetic energy in it. From where does it get this kinetic energy?

(c) What idea about work and force does the following diagram describe?

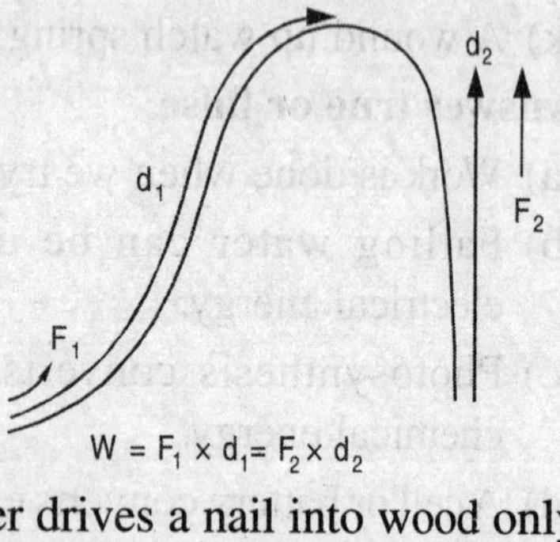


(d) Hammer drives a nail into wood only when lifted and then struck.

(e) In what way will the temperature of water at the bottom of a waterfall be different from the temperature at the top?

(f) A truck driver starts off his loaded truck. What are the major energy changes that take place in setting the truck in motion?

(g) In each of the following figures, a force '*F*' is acting on an object of mass '*m*'. The direction of displacement '*d*' is from left to right as shown. State whether the work done by the force is negative, positive or zero.

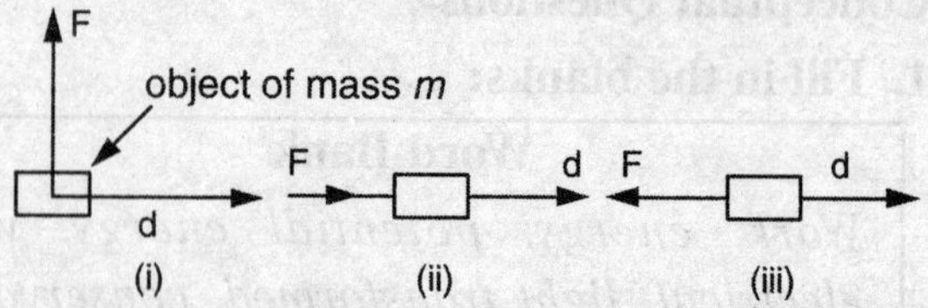


(h) A bullet is fired from a gun. Which will have greater kinetic energy, the bullet or the gun?

(i) In a tug of war, one team is slowly giving way to the other. What work is being done, and by whom ?

(j) What is the amount of work done when an object moves in a circular path for one complete rotation?

(k) Describe why chemical energy is a form of potential energy.

**9.** When you hit a nail into a board by using a hammer, the head of the nail gets warm. In terms of kinetic and thermal energy, describe why you think the nail head gets warm.

**10.** Explain why a high speed collision may cause more damage to vehicles than a low speed collision does.

**11.** How is elastic potential energy stored and released?

**12. State the energy changes taking place in the following:**

(a) An oscillating pendulum

(b) Photosynthesis in green leaves

(c) Charging of a battery

(d) The unwinding of a spring of a watch

(e) A dynamo (f) An electric toaster

(g) A photoelectric cell (h) An electromagnet

(i) A loudspeaker (j) An alarm clock

(k) A light bulb (l) A hair dryer

**13.** Describe the kinetic-potential energy conversion that occurs when a basketball bounces.

**14.** Brakes are suddenly applied to a speeding car and it comes to a screeching halt. Is the sound energy produced in this conversion a useful form of energy? Explain your answer.

**15.** (a) What kind of energy does the skier have at the top of the slope?
(b) What happens to that energy after the skier races down the slope of the mountain?

**16.** Describe what happens in terms of energy when you blow up a balloon and release it.

**17.** Imagine that the sun ran out of energy. What would happen to our energy resources on Earth?

**Multiple Choice Questions (Q. 18– Q. 36) :**

**18.** A pendulum is oscillating about its mean position in vacuum. It has
(a) only kinetic energy
(b) maximum kinetic energy at extreme position
(c) maximum potential energy at its mean position
(d) sum of kinetic energy and potential energy remains constant throughout the motion.

**19.** A stretched spring possesses _____ energy.
(a) kinetic (b) elastic potential
(c) electrical (d) magnetic

**20.** A raised hammer possesses
(a) K.E. only (b) gravitational P.E.
(c) electrical energy (d) none of these

**21.** In plants, energy is transformed from
(a) kinetic to potential (b) light to chemical
(c) chemical to electrical (d) chemical to light

**22.** Which of the following is a renewable source of energy ?
(a) wind (b) coal
(c) nuclear energy (d) petroleum

**23.** A bird flying in the sky has
(a) K.E. only (b) P.E. only
(c) neither K.E. nor P.E. (d) both K.E. and P.E.

**24.** The flowing of water possesses _____ energy.
(a) gravitational (b) kinetic
(c) potential (d) electrical

**25.** A steam engine converts
(a) heat energy into sound energy
(b) mechanical energy into heat energy
(c) heat energy into mechanical energy
(d) electrical energy into sound energy

**26.** When energy changes from one form to another, some of the energy always changes into
(a) kinetic energy (b) potential energy
(c) thermal energy (d) mechanical energy

**27.** Priyanka lifts a doll from the floor and places it on a table. If the weight of the doll is known, what else does one need to know in order to calculate the work done by Priyanka on the doll?
(a) The time taken in lifting the doll to the table
(b) Height of the table
(c) Mass of the ball
(d) Cost of the ball or the table

**28.** A body is dropped from a certain height from the ground. When it is half-way down, it possesses:
(a) only K.E. (b) zero energy
(c) only P.E. (d) both K.E. and P.E.

**29.** K.E. depends on
(a) mass and volume (b) velocity and weight
(c) weight and height (d) velocity and mass

**30.** Which of the following types of energy is/are not a renewable source?
(a) wind energy (b) nuclear energy
(c) solar energy (d) chemical energy

**31.** A car is moving along a straight level road with constant speed. Then
(a) the work done on the car is a measure of its gravitational P.E.
(b) the work done on the car is zero.
(c) the work done on the car cannot be found.
(d) the work done on the car is infinite.

**32.** Which of the following sentences describes a conversion from chemical energy to thermal energy ?

(a) Food is digested and used to regulate body temperature.
(b) Charcoal is burnt in a barbecue pit.
(c) Coal is burnt to produce steam
(d) All of the above

**33.** When you compress a coil spring, you do work on it. The elastic potential energy

(a) increases (b) decreases
(c) disappears (d) remains unchanged

**34.** Potential energy of your body is minimum when you

(a) are standing
(b) are sitting on a chair
(c) are sitting on the ground
(d) lie down on the ground

**35.** A truck and a car are moving on a smooth level road such that the K.E. associated with them is the same. Which one will cover greater distance when brakes are applied to them simultaneously?

(a) car
(b) both will cover the same distance
(c) truck
(d) none of these

**36.** Mechanically work done is equal to (symbols have their usual meanings)

(a) $W = F/d$ (b) $W = Fd$
(c) $W = F + d$ (d) $W = F - d$

**37. Numerical problems :**
**Calculate**

(i) Work done if using a force of 12 N, a cart is pushed through 8 m.
(ii) Work done if a stool weighs 25 N is lifted to a height of 80 cm.
(iii) K.E. of a car of mass 400 kg moving with a speed of 20 ms$^{-1}$.
(iv) P.E. of a body of mass 8 kg raised to a height of 5 m. ($g = 10$ ms$^{-2}$).
(v) (a) Increase in the P.E. of a boy of mass 50 kg when he moves from a height of 5 m to 8 m. ($g = 10$ ms$^{-2}$)
(b) Work done by the boy against gravity.
(vi) How fast should a man of 70 kg run so that his kinetic energy is 1715 J ?
(vii) Work done by a man of weight of 300 N in climbing the second floor of a building 7.0 m high.
(viii) Speed of a 1 kg mass having K.E. equal to 1 J.
(ix) Velocity of a body of mass 100 g having a K.E. of 20 J.
(x) The height to which an object of 0.5 kg will rise if 1 J of energy is applied do it.
(xi) The kinetic energy of a boy of mass 50 kg sitting in a car moving with a uniform velocity of 72 km/hr.
(xii) Work done in holding a 15 kg bag while waiting for a bus for 10 minutes.

**38.** A mass of 4 kg originally at rest is pulled by a string along a smooth horizontal surface. The force of 15 N is maintained until the mass undergoes a displacement of 20m at which stage the mass is allowed to keep moving at constant velocity.

(a) How much work is done by the force ?
(b) What would be the velocity of the mass at the end of 20 m ?

**39.** A ball of mass 500 g falls from a height of 4 m. Find the K.E. of the ball when it reaches the ground. (Take $g = 9.8$ m/s$^2$).

**40.** A body of mass 10 kg is dropped from a height of 20 m. (Take $g = 10$ m/s$^2$)

(i) Find its potential energy before it is dropped
(ii) Its kinetic energy when it is 8 m above the ground
(iii) Its kinetic energy when it hits the ground.

**41.** Use the figure below to find the following :

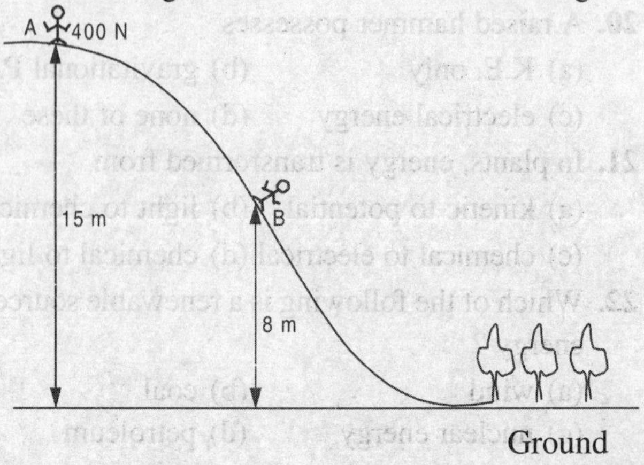

(a) the skier's gravitational potential energy at point *A*.

(b) the skier's gravitational potential energy at point *B*.

(c) the skier's kinetic energy at *B*.

**42.** A person lifts a parcel of mass 20 kg 1.5 m up from the ground and puts it in a truck. Find

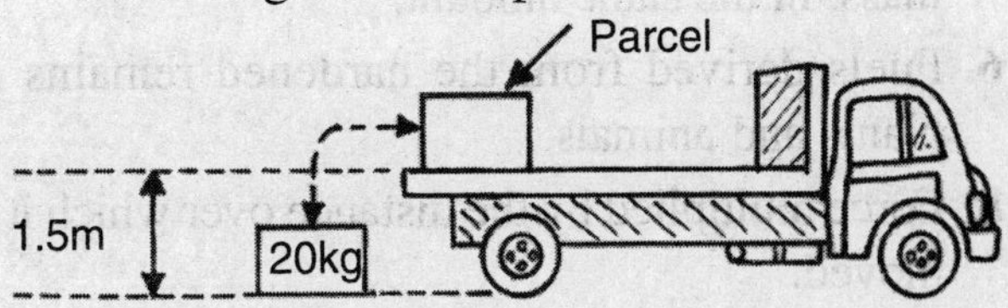


(a) What force needs to be applied to the parcel to move it up at constant speed ?

(b) What is the work done by the force ?

(c) The work is now stored as energy in the parcel. What type of energy is it ?

(d) If the parcel were to fall off the back of the truck, what would happen to this stored energy as it fell to the ground ?

**43.** Suppose that while riding bike, a person coasts down both a small hill and a large hill. Compare his final speed at the bottom of the small hill with his final speed at the bottom of the large hill. Explain your answer.

**44. Use the following terms to create a concept map.**

*Energy, Kinetic energy, Potential energy, Joule, Gravitational, Mechanical, Chemical, Electrical, Thermal, Sound, Light, Elastic Renewable, Non-renewable, Solar, Wind, Running water, Petrol, Coal, Fossil fuels, Biomass.*

**45. Solve the following crossword with the help of the given clues :**

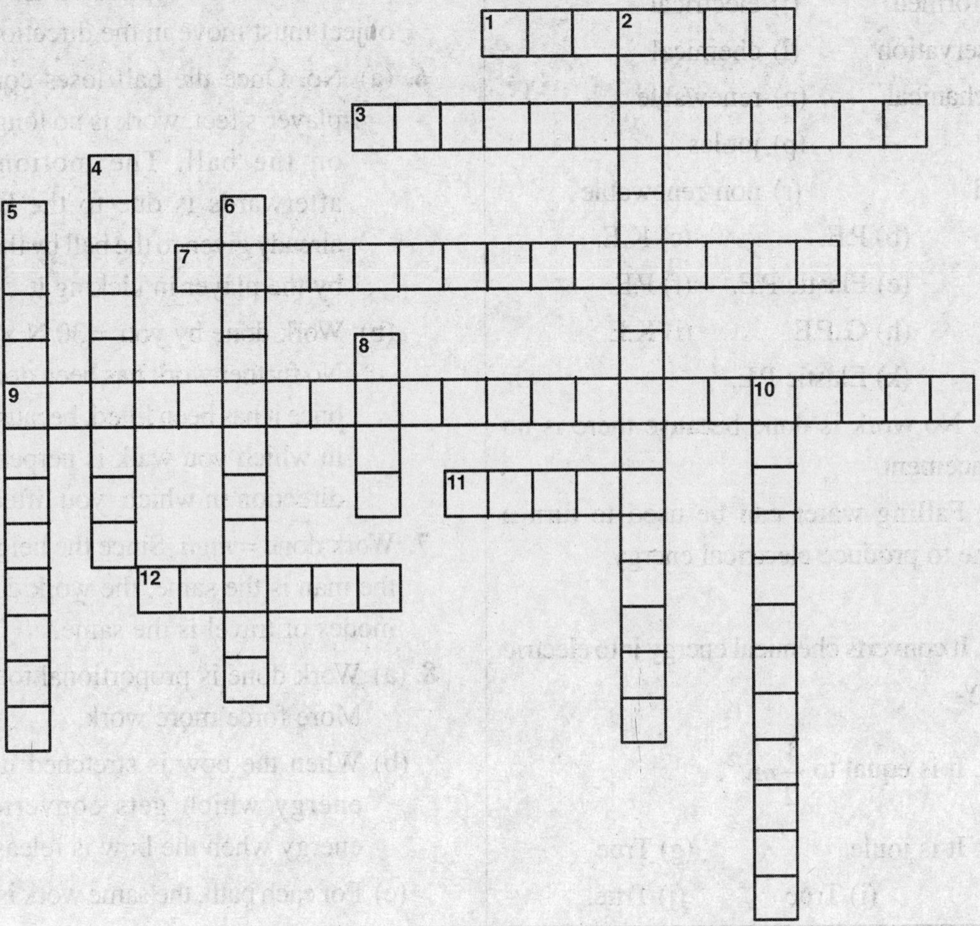

### ACROSS

1. Organic materials, such as plant matter and manure
3. The energy possessed by a body as a result of being in motion.
7. The energy possessed by a body as a result of its position or condition.
9. Energy stored when objects are stretched, compressed or bent.
11. Unit of work and energy
12. The capacity or power to do work.

### DOWN

2. Total energy of motion and position of an object.
4. Relating to a natural source, such as solar energy that is never used up or that can be replaced by new growth.
5. The continuance of a physical quantity, such as mass, in the same amount.
6. Fuels derived from the hardened remains of plants and animals.
8. Force multiplied by the distance over which it is moved.
10. Relating to a natural source that cannot be replaced once it has been used.

## Answers

**1.** (a) work (b) newton-metre, joule
(c) joule, metre (d) energy, joules
(e) potential energy (f) kinetic energy
(g) mechanical energy
(h) gravitational potential energy
(i) transformed (j) electrical
(k) conservation (l) chemical
(m) mechanical (n) renewable
(o) light (p) joules
(q) wind (r) non-renewable

**2.** (a) K.E. (b) P.E. (c) K.E.
(d) P.E. (e) Elastic P.E. (f) P.E.
(g) K.E. (h) G.P.E. (i) K.E.
(j) P.E. (k) Elastic P.E.

**3.** (a) False. No work is done because there is no displacement
(b) True. Falling water can be used to turn a turbine to produce electrical energy.
(c) True
(d) False. It converts chemical energy into electric energy.
(e) False. It is equal to $\frac{1}{2}mv^2$.
(f) False. It is joule. (g) True
(h) True (i) True (j) True.
(k) False. No work is done because the suitcase does not move.

**4.** Work occurs when a force causes an object to move in the direction of the force. Joule is the unit in which work is measured.

**5.** A force must be applied on an object and the object must move in the direction of the force.

**6.** (a) No. Once the ball loses contact with the player's feet, work is no longer being done on the ball. The motion of the ball afterwards is due to the kinetic energy already given to the ball by the force applied by the player in kicking it.
(b) Work done by you = 30 N × 0.8 m = 24 J; No further work has been done on the chair once it has been lifted, because the direction in which you walk is perpendicular to the direction in which you lifted the chair.

**7.** Work done = $mgh$. Since the height covered by the man is the same, the work done in the two modes of travel is the same.

**8.** (a) Work done is proportional to applied force. More force more work.
(b) When the bow is stretched it has potential energy which gets converted to kinetic energy when the bow is released.
(c) For each path, the same work is done to move

the object to the top of the hill, although distance and force along the two paths differ.

(d) Hammer when lifted acquires gravitational potential energy which is converted into kinetic energy.

(e) Temperature at the bottom is more, because the potential energy at the top changes into heat energy when water reaches the bottom.

(f) Chemical energy changes into heat energy and finally into kinetic energy that sets the truck into motion.

(g) (i) zero, because there is no displacement in the direction parallel to the force, (ii) positive, because displacement takes place in the direction of the force, (iii) negative, because displacement takes place in a direction opposite to the direction of the force.

(h) The bullet will have a greater K.E. because of its great speed.

(i) Work is done by the winning team and is equal to the product of the net force (difference of the forces applied by the two teams) and the displacement that the losing team suffers.

(j) Zero. When an object moves along a circular path, the force acting on it is normal (at right angles) to the direction of motion.

(k) It depends on the position of atoms in a molecule.

**9.** Kinetic energy is used to nail the hammer into the board, and that energy is converted into thermal energy.

**10.** A vehicle moving at high speed has a lot more kinetic energy than a vehicle moving at low speed has, so the high speed vehicle is able to cause more damage in case of a collision.

**11.** Elastic potential energy can be stored by stretching a rubber band. It is released when the rubber band goes back to its original shape.

**12.** (a) When a pendulum swings its speed at mean position is maximum and being the lowest point, height is 0. Hence, at mean position K.E. is maximum and P.E. is zero. When it swings to either side of the mean position, its height from the ground increases and speed decreases. So K.E. changes gradually into P.E. In the extreme positions, K.E. becomes zero and P.E. becomes maximum and the pendulum in these positions momentarily comes to rest.

(b) Light energy into chemical energy.

(c) Electrical energy into chemical energy.

(d) A wound up watch spring has elastic potential energy due to its wound up state. As the spring unwinds, the P.E. stored in it changes into kinetic energy which does work in moving the hands of the watch.

(e) Mechanical energy into electrical energy

(f) Electrical energy into heat energy

(g) Light energy into electrical energy

(h) Electrical energy into magnetic energy

(i) A loudspeaker when in use converts electrical energy, received in the form of electrical signals from the microphone, into sound energy.

(j) Electrical energy $\rightarrow$ light energy and sound energy.

(k) Electrical energy $\rightarrow$ light energy and thermal energy.

(l) Electrical energy $\rightarrow$ kinetic energy, thermal energy and sound energy.

**13.** The potential energy of a basketball is maximum when it reaches the greatest height. As it falls to the ground, the potential energy is gradually converted into kinetic energy. The K.E. is maximum when the ball reaches the ground. As the ball bounces after hitting the ground, the kinetic energy gradually reduces and gets converted into potential energy. This process continues.

**14.** Since the sound energy produced by the application of sudden brakes does not contribute to stopping the car, it is not a useful form of energy.

**15.** (a) Potential energy, (b) It is converted to K.E.

**16.** The compression of the air in the balloon is a kind of potential energy, which is released in the form of kinetic energy when the balloon is let gone.

**17.** There would no longer be any source of solar, wind or hydroelectric energy.

| | | | |
|---|---|---|---|
| **18.** (d) | **19.** (b) | **20.** (b) | **21.** (b) |
| **22.** (a) | **23.** (d) | **24.** (b) | **25.** (c) |
| **26.** (c) | **27.** (b) | **28.** (d) | **29.** (d) |
| **30.** (b), (d) | **31.** (b) | **32.** (d) | |

**33.** (a) The work done by you will be stored in the form of elastic potential energy.

**34.** (d) When you lie down on the ground the height of your centre mass is minimum, therefore, P.E. is minimum.

**35.** (b) **36.** (b)

**37.** (i) Work done = 12 N × 8 m = 96 J.

(ii) $80\text{ cm} = \frac{80}{100}\text{ m} = 0.8\text{ m}$.

Work done = 25 N × 0.8 m = 20 J

(iii) $\text{K.E.} = \frac{1}{2}mv^2 = \frac{1}{2}\times 400\text{ kg} \times (20\text{ m/s})^2$

$= 80000\text{ J}$

(iv) $\text{P.E.} = mgh = 8\text{ kg} \times 10\text{ ms}^{-2} \times 5\text{ m} = 400\text{ J}$.

(v) (a) Initial P.E. of the boy at 5 m height = $mgh$

$= 50\text{ kg} \times 10\text{ ms}^{-2} \times 5\text{ m} = 2500\text{ J}$

Final P.E. of the boy at 8 m height

$= 50\text{ kg} \times 10\text{ ms}^{-2} \times 8\text{ m} = 4000\text{ J}$

∴ Increase in P.E. = 4000 J – 2500 J = 1500 J.

(b) Work done by the boy against gravity

= Increase in P.E. = 1500 J.

(vi) $\text{K.E.} = \frac{1}{2}mv^2 \Rightarrow \frac{1}{2}\times 70\times v^2 = 1715$

$\Rightarrow v^2 = \frac{1715\times 2}{70} = 49 \Rightarrow v = 7\text{ m/s}$.

(vii) Work done = $F \times d$

= 300 N × 7 m = 2100 J.

(viii) $\text{K.E.} = \frac{1}{2}mv^2 \Rightarrow 1\text{J} = \frac{1}{2}\times 1\text{ kg}\times v^2$

$\Rightarrow v^2 = 2 \Rightarrow v = \sqrt{2}\text{ m/s} = 1.4\text{ m/s approx.}$

(ix) $20\text{ J} = \frac{1}{2}\times\frac{100}{1000}\text{kg}\times v^2 \Rightarrow v^2 = 20\times 20 = 400$

$\Rightarrow v = 20\text{ m/s}$.

(x) $\text{P.E.} = mgh \Rightarrow 1\text{ J} = 0.5\text{ kg}\times 10\times h \Rightarrow h = 0.2\text{ m}$.

(xi) Velocity of the boy = Velocity of the car = 72 km h

$= 72\times\frac{5}{18}\text{ms}^{-1} = 20\text{ms}^{-1}$.

$\therefore$ K.E. of the boy $= \frac{1}{2}mv^2 = \frac{1}{2}\times 50\text{ kg}\times(20\text{ ms}^{-1})^2$

= 10000 J.

(xii) Work done is zero because displacement is zero.

**38.** (a) Work done = Force × displacement

= 15 N × 20 m

= 300 J

(b) $\text{K.E.} = \frac{1}{2}mv^2 = 300\text{ J}$

$\Rightarrow \frac{1}{2}\times 4\times v^2 = 300$

$\Rightarrow v^2 = 150 \Rightarrow v = \sqrt{150} = 12.25\text{ ms}^{-1}$.

**39.** $\text{Mass} = 500\text{ g} = \frac{500}{1000}\text{kg} = \frac{1}{2}\text{kg}$

P.E. at a height of 4 m = $mgh$

$= \frac{1}{2}\text{ kg}\times 9.8\text{ ms}^{-2}\times 4\text{ m} = 19.6\text{ J}$

P.E. of the ball when it reaches the gorund = 0

∴ K.E. of the ball when it reaches the ground = Loss in P.E. = 19.6 J.

**40.** (i) $\text{P.E.} = 10\text{ kg}\times 10\text{ ms}^{-2}\times 20\text{ m} = 2000\text{ J}$

(ii) P.E. at 8 m from the ground

$= 10\text{ kg}\times 10\text{ ms}^{-2}\times 8\text{ m} = 800\text{ J}$.

∴ K.E. at 8 m height = Change in P.E.

= 2000 J – 800 J = 1200 J

(iii) P.E. when it hits the ground = 0

∴ Kinetic energy when it hits the ground = Initial P.E. = 2000 J.

**41.** (a) 400 N × 15 m = 6,000 J

(b) 400 N × 8 m = 3,200 J

(c) 6,000 J – 3,200 J = 2,800 J.

**42.** (a) Reqd force = $mg$

$= 20\text{ kg}\times 9.8\text{ ms}^{-2}$

= 196 N

(b) Work done = $F \times d$

= 196 N × 1.5 m = 294 J

(c) Gravitational potential energy

(d) G.P.E would convert back to K.E. as the velocity increases.

**43.** The person would have a greater final speed at the bottom of the large hill than at the bottom of the small hill - the amount of gravitational potential energy depends on the height of the starting position. Starting from a greater height means starting with more gravitational potential energy, which is converted into kinetic energy as the person coasts down the hill.

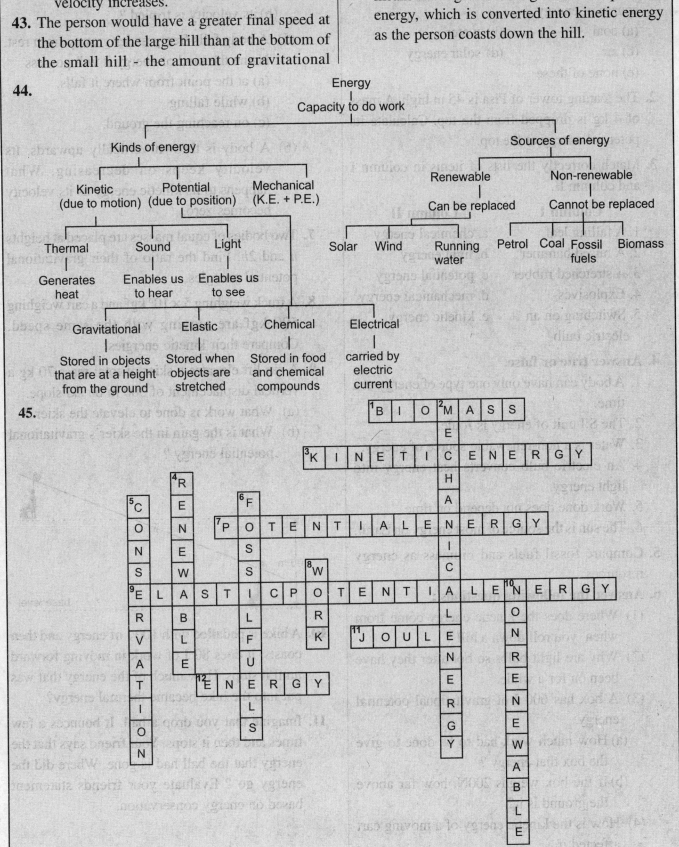

**44.**

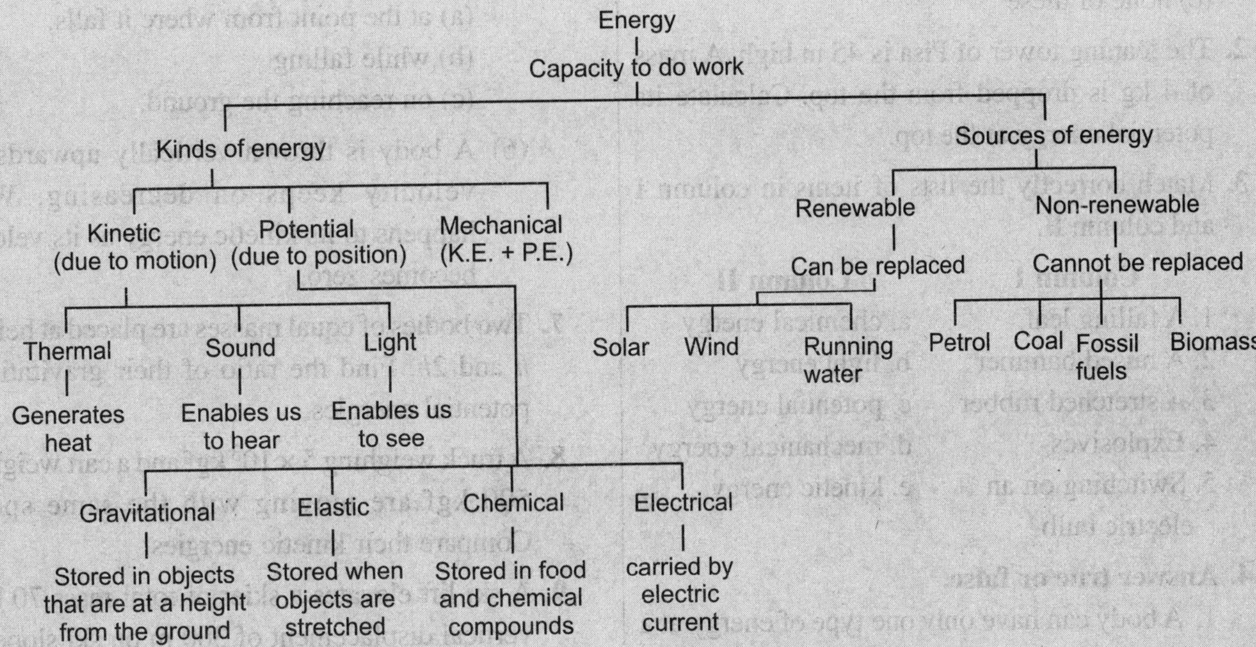


**45.**

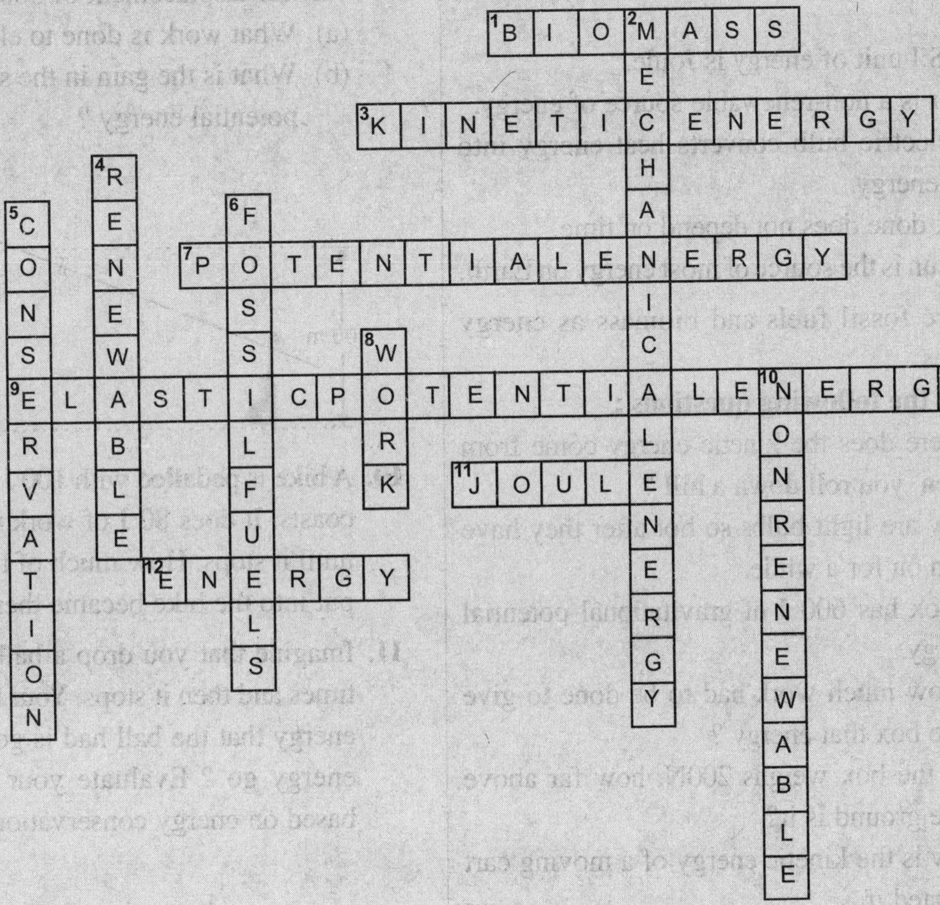

## Self Assessment Sheet-3

**1.** Which of the following is a non-renewable source of energy ?

(a) coal (b) water
(c) air (d) solar energy
(e) none of these

**2.** The leaning tower of Pisa is 45 m high. A mass of 4 kg is dropped from the top. Calculate its potential energy at the top.

**3.** Match correctly the lists of items in column I and column II.

| Column I | Column II |
|---|---|
| 1. A falling leaf. | a. chemical energy |
| 2. A raised hammer | b. light energy |
| 3. A stretched rubber | c. potential energy |
| 4. Explosives | d. mechanical energy |
| 5. Switching on an electric bulb | e. kinetic energy |

**4. Answer true or false.**

1. A body can have only one type of energy at a time.
2. The S.I unit of energy is Joule.
3. Water is a non-renewable source of energy.
4. An electric bulb converts heat energy into light energy.
5. Work done does not depend on time.
6. The sun is the source of most energy on Earth.

**5.** Compare fossil fuels and biomass as energy resources.

**6. Answer the following questions :**

(1) Where does the kinetic energy come from when you roll down a hill ?

(2) Why are light bulbs so hot after they have been on for a while.

(3) A box has 600 J of gravitational potential energy.

(a) How much work had to be done to give the box that energy ?

(b) If the box weighs 200N, how far above the ground is it?

(4) How is the kinetic energy of a moving cart affected if

(a) its mass is doubled,

(b) its velocity is tripled ?

(5) A body falls freely under gravity from rest. Name the kind of energy it will possess

(a) at the point from where it falls,

(b) while falling

(c) on reaching the ground.

(6) A body is thrown vertically upwards. Its velocity keeps on decreasing. What happens to its kinetic energy as its velocity becomes zero.

**7.** Two bodies of equal masses are placed at heights $h$ and $2h$. Find the ratio of their gravitational potential energies.

**8.** A truck weighing $5 \times 10^3$ kgf and a cart weighing 500 kgf are moving with the same speed. Compare their kinetic energies.

**9.** A ski lift elevates a skier of total mass 70 kg a vertical displacement of 300 m on ski slope.

(a) What work is done to elevate the skier ?

(b) What is the gain in the skier's gravitational potential energy ?

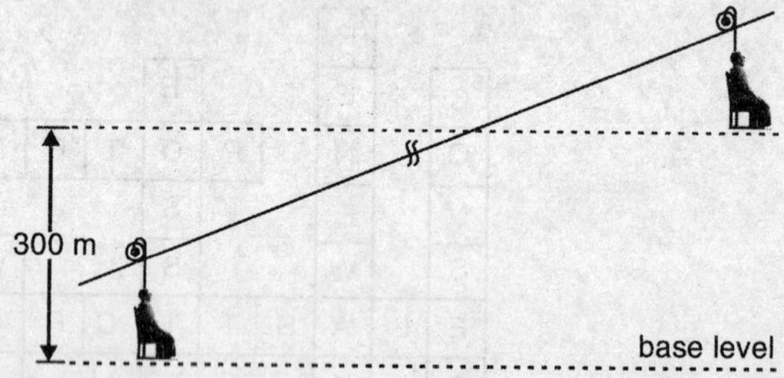


**10.** A bike is pedalled with 100 J of energy and then coasts. It does 80 J of work in moving forward until it stops. How much of the energy that was put into the bike became thermal energy?

**11.** Imagine that you drop a ball. It bounces a few times and then it stops. Your friend says that the energy that the ball had is gone. Where did the energy go ? Evaluate your friends statement based on energy conservation.

## Answers

**1.** a **2.** 1800 J

**3.** (1) – e, (2) – c, (3) –d or c, (4) –a, (5) – b

**4.** (1) False (2) True (3) False. It is a renewable source of energy. (4) False. An electric bulb converts electrical energy into heat and light energy. (5) True (6) True

**5.** Both fossil fuels and biomass come from organic matter, are burnt to generate energy, and cause air pollution when used. Biomass is renewable whereas fossil fuels are non-renewable.

**6.** (1) The kinetic energy comes from the potential energy that you had at the top of the hill.

(2) Because of the total energy, only one-tenth is converted to light energy. The rest is converted to thermal energy.

(3) (a) 600 J (b) 600 J ÷ 200 N = 3 m.

(4) K.E = $\frac{1}{2}$ $mv^2$, *i.e.*, it is directly proportional to the mass and proportional to square of the velocity. Therefore, if m is doubled (keeping the speed same), then kinetic energy is also doubled. If velocity is tripled (keeping the mass same) then the kinetic energy increases $3^2$, *i.e.*, 9 times.

(5) (a) Potential energy, (b) Potential energy and kinetic energy (c) Kinetic energy

(6) Kinetic energy changes to potential energy.

**7.** 1 : 2

**8.** Since K.E. ∝ mass, therefore,

$$\frac{(K.E.)_1}{(K.E.)_2} = \frac{m_1}{m_2} = \frac{5\times10^3}{500} = \frac{10}{1}$$

**9.** (a) Work done = $mgh$ = (70 kg × 10 ms$^{-2}$ × 300 m)
= 210000 J
= $2.1 \times 10^5$ J

(b) $2.1 \times 10^5$ J

**10.** 100 J – 80 J = 20 J

**11.** The energy that the ball had initially is not 'gone' but was converted into thermal energy and sound energy.

Chapter 4

# LIGHT

## KEY FACTS

1. Light is a form of energy.
2. The objects like sun and other stars that give out or emit light of their own are called **luminous** objects. The objects like moon, wall, air, water etc., which do not give light of their own are called **non-luminous** objects.
3. The production or emission of light by a living organism as a result of biochemical reaction within the body and conversion of chemical energy into light energy is called **Bioluminescence**. Fireflies are insects that give off a pale, greenish yellow light that flashes or glows in the dark. They are bioluminescent. Some of the fish, such as angler fish, living deep under the sea are also bioluminescent.
4. Substances like gems, ivory and paraffin that continue to emit light even after the source of light is withdrawn are known as **phosphorescent** substances.
5. Substances that give off light energy only as long as they receive some radiant energy are called **fluorescent** substances.
6. Objects may be ***transparent***, ***translucent*** **or** ***opaque***.
   (i) Objects like glass through which light rays can pass easily and through which we can see clearly are called **transparent objects**.
   (ii) Objects like oil sheets, muddy water through which only a part of the light can pass and it is difficult to see through them are called **translucent objects**.
   (iii) Objects like a wall and metal sheets through which no light can pass and we cannot see through them are called **opaque objects**.
   (iv) The light emitted by a luminous body is spread in all directions. This phenomenon is called **radiation**. We say that light radiates from the body. The rays of light coming out from sun are called **solar radiations**.

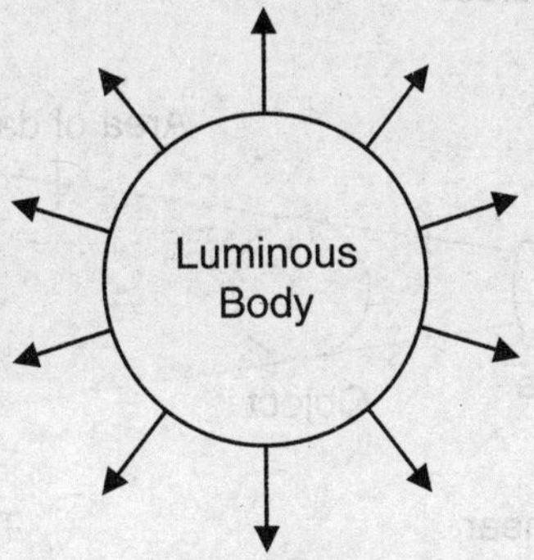


The luminous body spreads light in all directions.

   (v) Light always travels in straight lines unless it meets something that makes it change its direction. The property of light travelling in straight lines in a medium is called ***rectilinear propagation of light.***

Because of motion in straight lines, light rays are represented by rays.

**A ray of light is a path along which light energy travels.**

6.

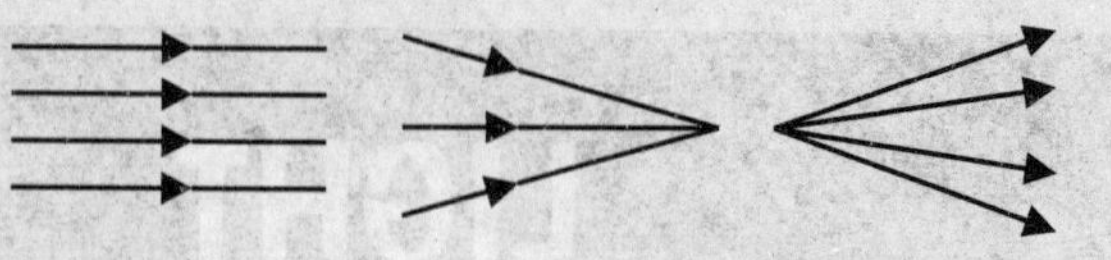


(i) Fig. (a) shows **parallel beam** of light. A search light emits a parallel beam of light. The rays of light from very distant objects like sun are considered to be parallel.

(ii) Fig. (b) shows a **convergent beam** of light. The light rays coming from different directions converge at a point.

**Ex.** A source behind a lens in a projection lantern can provide a convergent beam.

(iii) Fig. (c) shows a **divergent beam** of light. The light rays coming out from a source diverge in different directions.

**Ex.** A lamp emits a divergent beam of light. A candle flame also sends out rays in all directions.

**7. Shadow formation and eclipses.**

Shadows are formed when the path of the light is obstructed by an opaque object. The formation of shadows has the following characteristics :

(a) Shadow is formed in the direction opposite to the side of the light source.

(b) Brighter the light, darker is the shadow.

(c) The size of the shadow depends on the size of the light source and object, and also on the distance between the light source and the object, the wall or screen.

When the obstacle is near the screen the shadow formed is small and when the distance between the obstacle and the screen is increased the shadow becomes bigger and bigger.

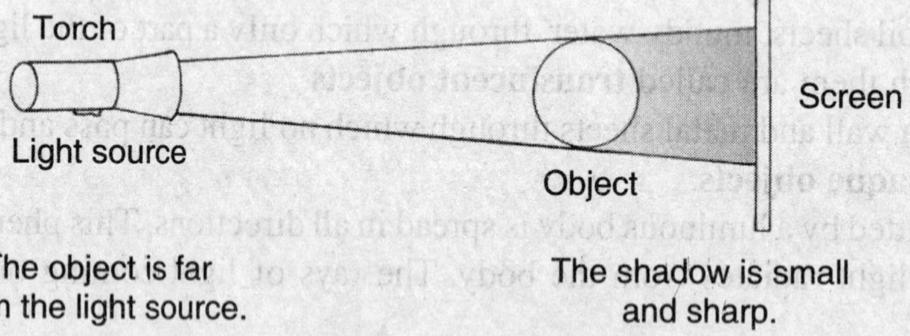


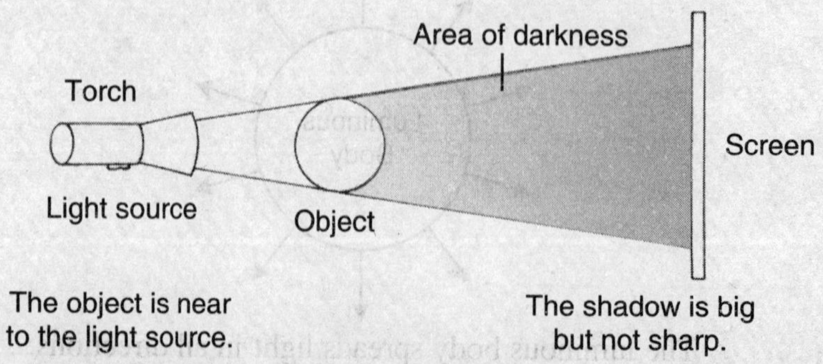


(d) The shape of a shadow depends on the shape of the object. An object can have shadows of different shapes, depending on where the light comes from.

The shadow of a cup looks as shown when light falls on the
(i) side, (ii) top or bottom, (iii) side with the handle.

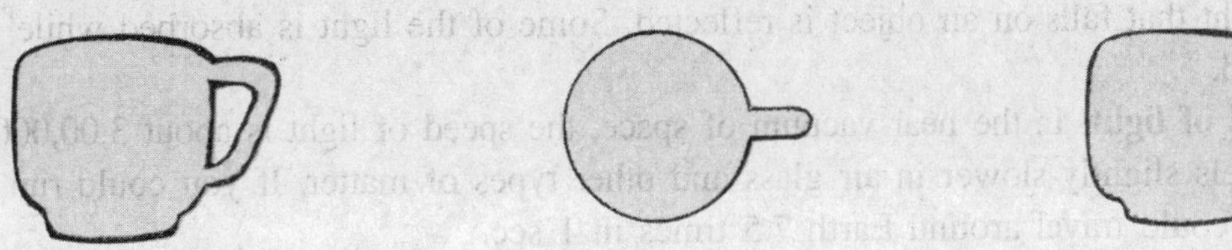

(i) side (ii) top or bottom (iii) side with the handle

(e) **Shadows cast by the sun** change in length and position at different times of the day. The shadows are long when the sun is low in the sky (morning). They are short when the sun is high in the sky (noon and afternoon). The sun is lowest in the sky at sunrise and at sunset, and highest at noon.

**8. Umbra and Penumbra**. The central darkest part of a shadow is called ***umbra***. The dark, central part of the shadow that is cast by the Moon onto the Earth during a solar eclipse is the umbra. The surrounding partially dark region is called ***penumbra***. The lighter, outer part of the shadow that is cast by the Moon onto the Earth during a solar eclipse is the penumbra.
If the screen is moved away from the obstacle it will be found that the penumbra goes on increasing and umbra gradually fades away.
**Note :** If your eye is in the region of umbra, no part of this source is visible and if your eye is in the region of penumbra, only a part of source is visible.

**9. Eclipse** is the partial or total blocking of light from one object by another object in outer space. During an eclipse, one object comes in between two other objects. In an eclipse of the Sun, the Moon passes between the Sun and the Earth. This is called ***Solar eclipse***. In case of eclipse of the moon, bright full moon is shadowed by the earth. This is called ***lunar eclipse***.

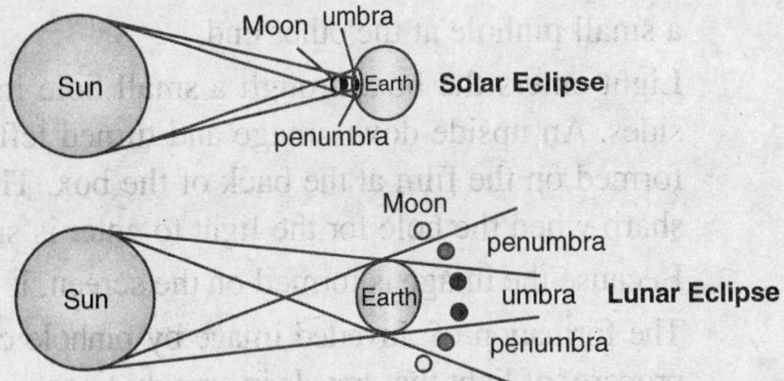


**10. Reflection** (1) The throwing back of a light wave, when it runs into an obstacle, such as a mirror or wall is called ***reflection.*** The reflection of light is just like a ball hitting a wall and bouncing off it. However, unlike the ball, light does not slow down each time it is reflected, although some of it may be absorbed.

(2) The non-luminous objects can be seen only when (i) light from some luminous object like the sun, electric bulb, torch light falls on them and (ii) it is reflected into our eyes. Thus, tree or a wall does not have its own light. We are able to see it when sunlight falls on it and it reflects the falling sunlight into our eyes.

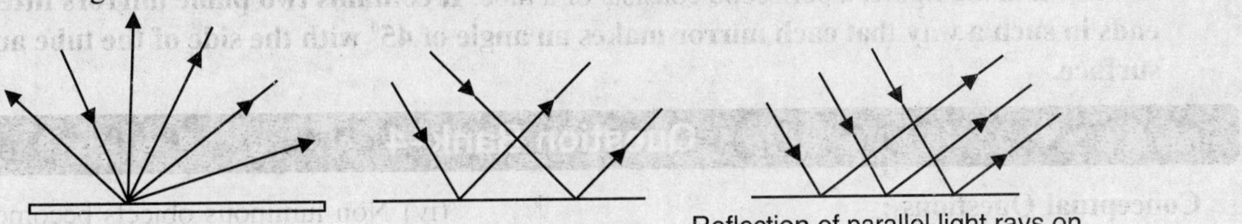

Irregular reflection of light from a rough surface

Regular reflection from a surface

Reflection of parallel light rays on a smooth surface. The reflected light rays are parallel.

(3) The amount and direction of the reflected light depends on the nature of the surface it strikes. Objects having rough surfaces reflect light in all directions. This is called ***irregular reflection***. For example, light falling on a piece of paper which has a rough surface is reflected in all directions. It is due to the irregular reflection of light that we cannot see the image of our face on looking into a piece of paper. Mirrors and polished metals have smooth surfaces and reflect all the light in a particular direction. This is called

***regular reflection***. It can form an image. Thus, the regular reflection of light by the smooth surface of a plane mirror produces an image of our face when we look into the mirror.

(4) Not all light that falls on an object is reflected. Some of the light is absorbed while the rest may be transmitted.

**11. The speed of light**. In the near vacuum of space, the speed of light is about 3,00,000 (3 lakh) km/s. Light travels slightly slower in air glass and other types of matter. If you could run at the speed of light, you could travel around Earth 7.5 times in 1 sec.

**12.** Earth is 15,00,00,000 (15 crore) km away from the sun, so it takes about 8.3 min for light to travel from the Sun to Earth

$$\text{Time} = \frac{\text{Distance}}{\text{Speed}}$$

$$= \frac{15,00,00,000 \text{ km}}{3,00,000 \text{ km/sec}}$$

$$= 500 \text{ sec} = \frac{500}{60} \text{ min}$$

$$= 8.3 \text{ min.}$$

**13. Pinhole Camera.** Also called a box camera, it is a simple device which forms image of a bright object on screen and consists of a closed box with a screen at one end and a small pinhole at the other end.

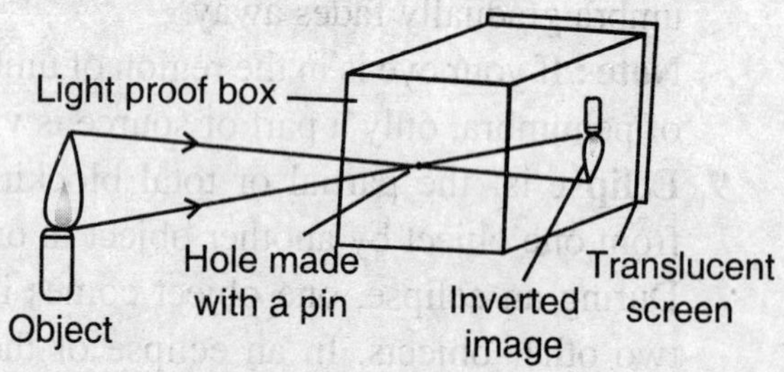


Light enters the box through a small hole in one of its sides. An upside down image and turned left-to-right is formed on the film at the back of the box. The image is sharp when the hole for the light to enter is small.

Because the image is formed on the screen, it is called a real image. The formation of inverted image by pinhole camera is based on the property of light that travels in straight lines.

**14. Periscope.** A periscope is a device by which it is possible to see an object on the other side of a high wall. It also enables soldier sitting in a trench observe the movements of the enemy above the ground and also a person sitting in a submarine submerged in water see the ships over the surface of the sea.

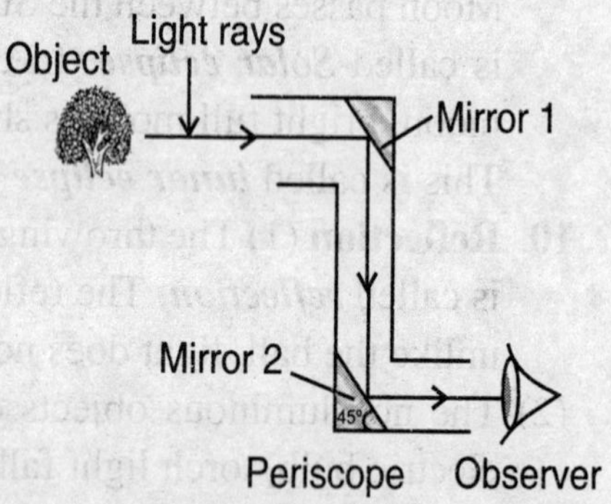


As shown in the figure, a periscope consists of a tube. **It contains two plane mirrors fitted at the two ends in such a way that each mirror makes an angle of 45° with the side of the tube and reflecting surface.**

## Question Bank–4

**Conceptual Questions :**

**1. What is the missing word?**

(i) Light travels in a ________.

(ii) Objects which emit light themselves are called ________ bodies.

(iii) Bodies which do not emit any light are called ________ bodies.

(iv) Non-luminous objects become ________ when they reflect light into our eyes.

(v) Substances through which light cannot pass are said to be ________.

(vi) The central and darker portion of the shadow is known as ________.

(vii) The image which cannot be projected on to a screen is called ________ image.

(viii) Substances that continue to emit light even after the source of light illuminating them is withdrawn are known as _________ substances.

(ix) Substances that stop giving off light once the source of light illuminating them is removed, are called _________ substances.

**2. Match the following:**

| | |
|---|---|
| (i) Stars | (a) Translucent |
| (ii) Human beings | (b) Transparent |
| (iii) Air | (c) Opaque |
| (iv) Torch light | (d) Natural sources of light |
| (v) Frosted glass | (e) Artificial source of light |

3. Which of the following is a luminous body?
(a) earth (b) moon
(c) sun (d) none of these

4. Objects which allow some light to pass through them but through which objects cannot be seen are called
(a) opaque (b) transparent
(c) translucent (d) none of these

5. Which of the following is a non-luminous body?
(a) fire (b) sun
(c) stars (d) earth

6. Which of the following is a luminous body?
(a) Fire (b) Earth
(c) Moon (d) Tree

7. The path along which light travels in a homogeneous medium is called the
(a) beam of light (b) ray of light
(c) pencil of light (d) none of these

8. A thin layer of water is transparent but a vey thick layer of water is
(a) translucent (b) opaque
(c) most transparent (d) none of these

**9. State whether the following statements are true or false:**

(i) A translucent object stops light only partially.

(ii) Birds do not cast shadows when they fly at a great height.

(iii) The moon is a luminous object.

(iv) A red object casts a red shadow.

(v) The colour of the shadow is black if the colour of the object is white.

(vi) An opaque object stops the light completely, therefore, it casts a dark shadow.

(vii) Penumbra is darker than umbra.

**10. Name the following:**

(i) The eclipse that takes place when moon comes between sun and earth.

(ii) The eclipse that takes place when earth comes between sun and moon.

11. Match the headers with the figures:

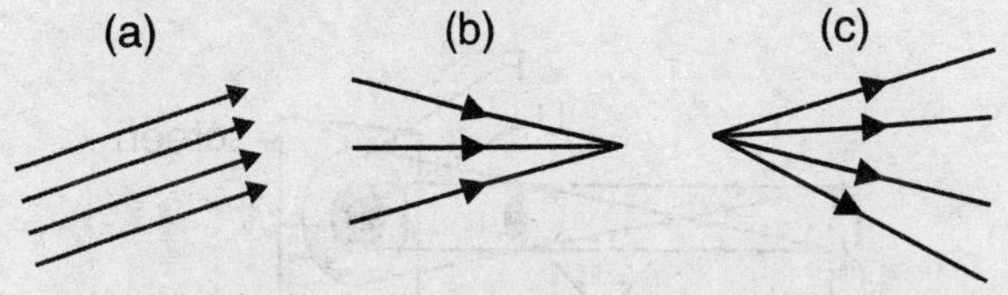


(1) Divergent (2) Parallel (3) Convergent

12. How long, to the nearest minute, does it take for light from the Sun to reach the Earth?
(a) 7 minutes (b) 8.3 minutes
(c) 8 minutes (d) 7.9 minutes

13. What is the missing word?
This experiment demonstrates that light travels in ------- lines.

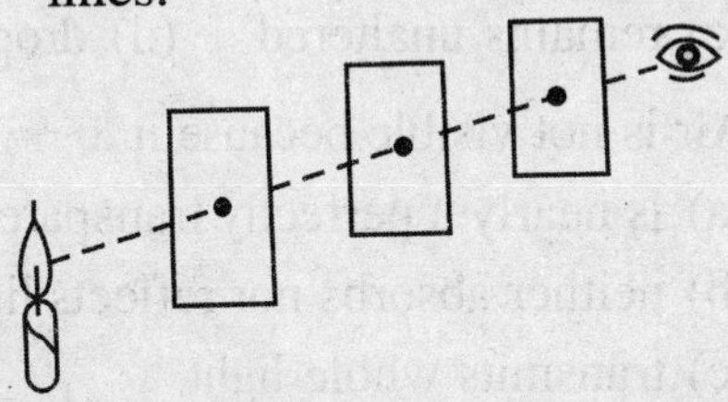

14. You can see through a car window because the window is
(a) opaque
(b) transparent
(c) translucent

15. The distance from Earth to the moon is 3,84,000 km. Calculate the time it takes for light to travel that distance.

16. Which of the following bodies allows only a part of the light to pass through it?
(a) Oiled paper (b) Brick
(c) Wood (d) Air

**17.** Rearrange the boxes given below to make a sequence that helps us understand opaque objects.

OWS | AKE | OPAQ | UEO | BJEC

TSM | SHAD

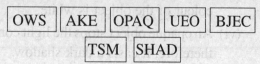

**18.** Can you think of creating a shape that would give a circular shadow if held in one way and a rectangular shadow if held in another way?

**19.** In a completely dark room if you hold up a mirror in front of you, will you see a reflection of yourself in the mirror ?

**20.** What is the missing label at (i) P, (ii) U

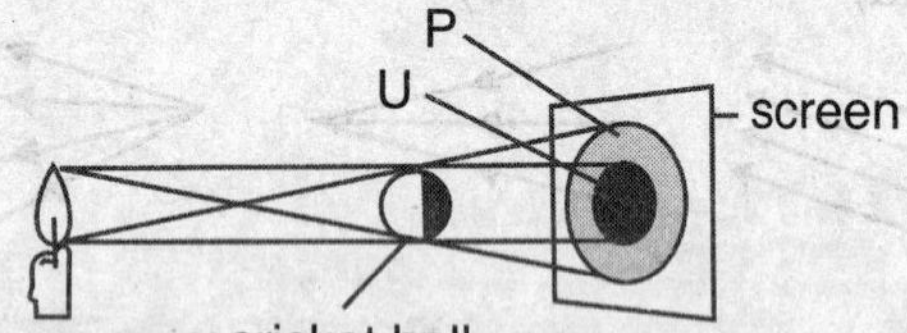


**21.** A substance which radiates light when heated to a high temperature is said to be

(a) luminescent (b) incandescent
(c) flourescent (d) phosphorescent

**22.** The speed of light with the rise in the temperature of the medium

(a) increases (b) decreases
(c) remains unaltered (d) drops suddenly

**23.** Air is not visible because it

(a) is nearly a perfectly transparent substance
(b) neither absorbs nor reflects light
(c) transmits whole light
(d) all the above are correct

**24.** We are able to see things because

(a) the light emitted by the objects reaches our eyes
(b) the light emitted by our eyes falls on the objects
(c) the light scattered by the objects reaches our eyes
(d) the rays of sun fall on the objects

**25.** Lunar eclipse takes place on

(a) new moon (b) full moon
(c) 22nd of the month (d) 8th of the month

**26.** To an astronaut the outer space appears

(a) white (b) black
(c) deep blue (d) crimson

**27.** Circular shadow of Earth on the Moon occurs during

(a) lunar eclipse (b) solar eclipse
(c) sunrise (d) sunset

**28.** If there were no atmosphere, what would be the colour of the Earth?

(a) red (b) blue
(c) black (d) white

**29.** The solar eclipse achieves totality only in limited geographical region because

(a) The trajectories of the earth around the sun and the moon around the earth are not perfect circles
(b) The size of the shadow of moon on the earth is small compared to the cross-section of the earth
(c) The earth is not a smooth flat surface, but has elevations and depressions

**30.** Who measured the speed of light first?

(a) Galileo (b) Newton
(c) Romer (d) Einstein

**31.** The amount of light reflected depends upon

(a) the colour of material of the object
(b) the nature of the surface
(c) the smoothness of the surface
(d) all the above

**32.** The object shown at the right is taken out in the sun and rotated. The different shadows formed are observed. Mark the picture which cannot be the shadow of this object.

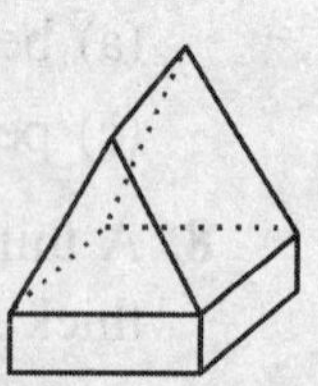

(a)  (b) 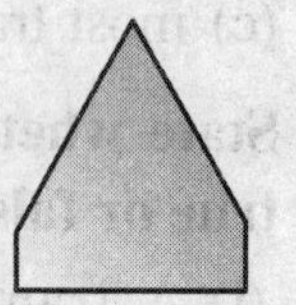

(c) 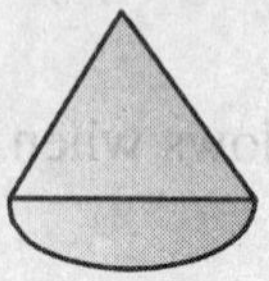 (d) 

**33. Solve the following crossword with the help of the given clues.**

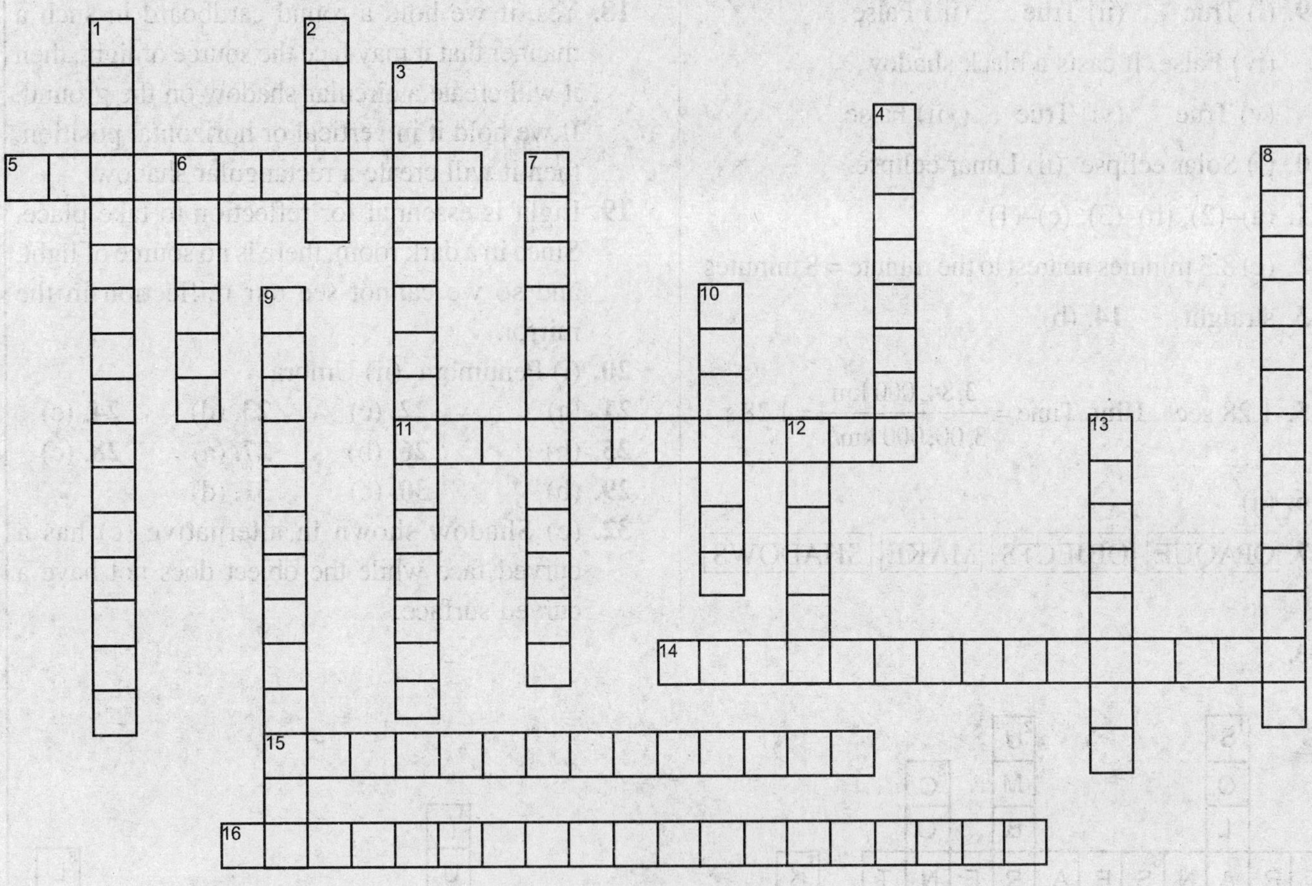


**ACROSS**

5. Transmitting light so as to be seen through clearly.
11. Objects that do not emit light.
14. Rays of light that diverge in different directions.
15. Substances that continue to emit light even after withdrawal of the source of light.
16. Throwing back of a light wave when it strikes an obstacle.

**DOWN**

1. Rays of light coming out from the sun.
2. The central and dark portion of a shadow.
3. The light rays that converge at a point.
4. Light emitting objects.
6. Formed when the path of the light is blocked by an opaque object.
7. Path along which light energy travels .
8. Eclipse of the moon.
9. Eclipse of the sun.
10. The blocking of light from one heavenly body by another.
12. Objects that do not allow light to pass through.
13. A region of partial shadow surrounding the umbra.

## Answers

**1.** (i) straight line (ii) luminous
(iii) non-luminous (iv) visible
(v) opaque (vi) umbra
(vii) virtual (viii) phosphorescent
(ix) fluorescent

**2.** (i) – (d) (ii) – (c) (iii) – (b) (iv) – (e) (v) – (a)

**3.** (c) **4.** (c) **5.** (d) **6.** (a)

**7.** (b) **8.** (a)

**9.** (i) True (ii) True (iii) False

(iv) False. It casts a black shadow.

(v) True (vi) True (vii) False

**10.** (i) Solar eclipse (ii) Lunar eclipse

**11.** (a)–(2), (b)–(3), (c)–(1)

**12.** (c) 8.3 minutes nearest to the minute = 8 minutes

**13.** straight **14.** (b)

**15.** 1.28 sec **Hint.** Time $= \dfrac{3,84,000 \text{ km}}{3,00,000 \text{ km/s}} = 1.28$ s

**16.** (a)

**17.** OPAQUE OBJECTS MAKE SHADOWS

**18.** Yes, if we hold a round cardboard in such a manner that it may face the source of light, then it will create a circular shadow on the ground. If we hold it in vertical or horizontal position, then it will create a rectangular shadow.

**19.** Light is essential for reflection to take place. Since in a dark room, there is no source of light, and so we cannot see our relflection in the mirror.

**20.** (i) Penumbra (ii) Umbra

**21.** (a) **22.** (c) **23.** (d) **24.** (c)

**25.** (b) **26.** (b) **27.** (a) **28.** (c)

**29.** (b) **30.** (c) **31.** (d)

**32.** (c) Shadow shown in alternative (c) has a curved face while the object does not have a curved surface.

**33.**

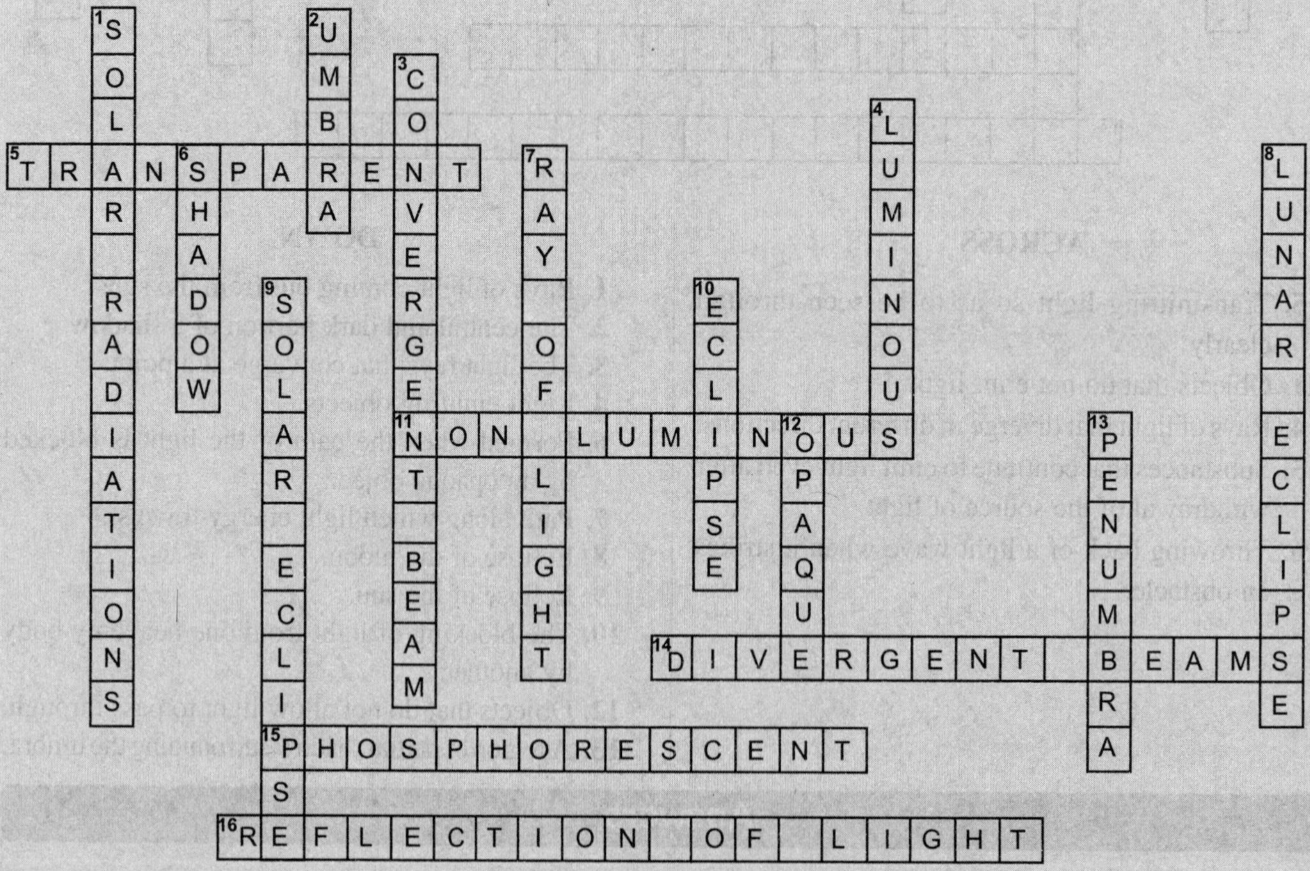

## Self Assessment Sheet-4

**1. Answer true or false :**

(i) Penumbra is a region of partial darkness.
(ii) Shadows are long when sun is overhead.
(iii) An wooden plank is an opaque object
(iv) Umbra is a region of complete darkness.
(v) Lunar eclipse occurs on new moon.
(vi) Objects which emit their own light are called non-luminous objects.
(vii) No shadow is formed when light passes through an object.

**2. Match correctly :**

| Column A | Column B |
|---|---|
| 1. Moon comes between sun and earth | a. non luminous |
| 2. Stars | b. natural source of light |
| 3. Allows light to pass through | c. luminous |
| 4. Earth | d. artificial source of light |
| 5. Candle | e. solar eclipse |
| 6. Sun | f. transparent |

**3. Give reasons for the following** :

(a) If there is a glass door and is closed then you may not be able to see it and get hurt if you try to pass through it.

(b) Although the moon is non-luminous, it appears bright and brings light on earth

(c) Silvered surfaces are used behind light bulbs in torches.

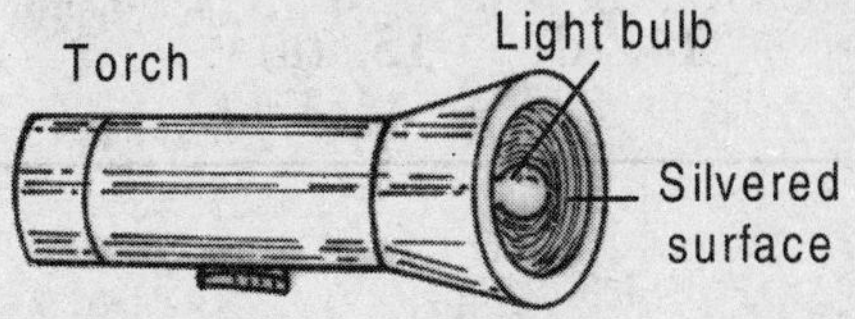


(d) If you pass under a tree covered with large number of leaves, you may notice small patches of light under it.

**4.** Put the following objects correctly in the given Venn diagram.
cardboard, air, glass, tracing paper, plastic scale, muddy water, brick, polythene, earth.

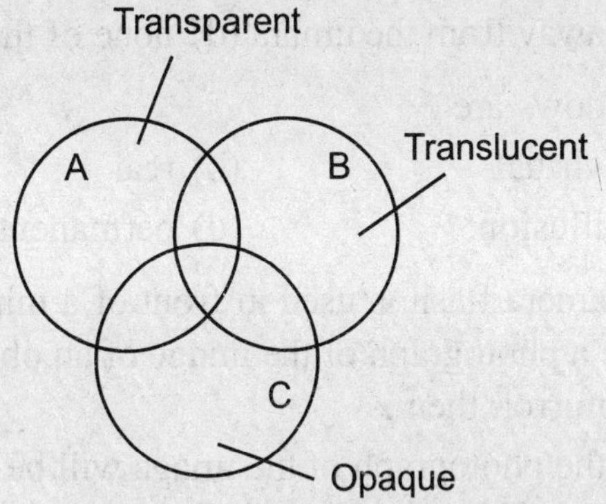


**MCQ'S. Tick the correct answer :**

**5.** Shadows are formed by
(a) light passing through an object.
(b) an opaque object blocking the path of light
(c) by light reflecting from an object
(d) by light deflecting from an object

**6.** As the object moves closer to a light source, the shadow
(a) gets bigger (b) gets smaller
(c) stays the same size

**7.** What kind of material is tissue paper ?
(a) trasparent (b) opaque
(c) translucent

**8.** We are able to see moon because
(a) it emits light (b) it absorbs light
(c) it reflects sunlight (d) it is luminous

**9. Which of the following are correct statements.**
An opaque object
(a) allows light to pass through it.
(b) does not allow light to pass through it.
(c) emits light.
(d) reflects light falling through it.
**A.** a, c **B.** b, c, d
**C.** b, d **D.** a, d

**10.** It is not possible to see a burning candle by using a bent pipe because

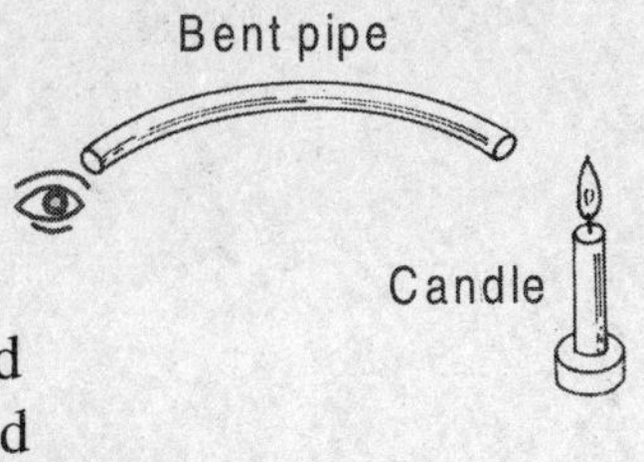


(a) light gets reflected
(b) light gets absorbed
(c) light travels in straight line
(d) light gets bent along the pipe.

**11.** During shadow formation, penumbra is seen
(a) inside the umbra (b) outside the umbra
(c) away from the umbra (d) none of these

**12.** Shadows are
(a) virtual (b) real
(c) illusion (d) permanent

**13.** A camera flash is used in front of a mirror to take a photograph of the image of an object in the mirror, then :
(a) the photograph of the image will be clearer
(b) the photograph of the image will be brighter
(c) the photograph of the object will be more beautiful
(d) the photograph of the image will not show anything except white colour.

**14.** A periscope in a submarine helps in viewing those objects :
(a) that are very far
(b) that do not emit any light
(c) that are below the surface of water
(d) that are above the surface of water

**15.** The light from torch when falls on a key, form a shadow on the screen behind the card as shown in the figure. As the torch is moved away from the card the shadow will

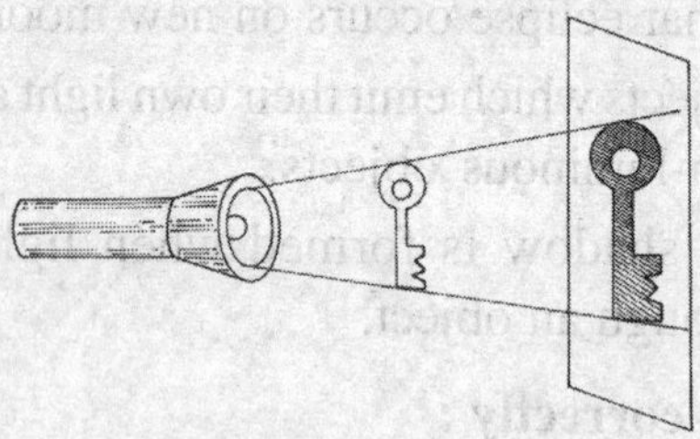

(a) become smaller
(b) become larger
(c) change its shape
(d) remain unchanged.

## Answers

**1.** (i) True (ii) False (iii) True (iv) True
(v) False. It occurs on full moon.
(vi) False (vii) True

**2.** 1 – e, 2 – c, 3 – f, 4 – a, 5 – d, 6 – b

**3.** (i) We are able to see an object only if the light falling on it is reflected to our eyes. A glass door being transparent allows most of the light pass through it and so it is very difficult to see the door.

(ii) Moon though non-luminous itself appears bright because being opaque the light falling on it from the sun is reflected to the earth.

(iii) The silvered surface, reflects out the light that falls on it.

(iv) The gaps between the leaves act as semi holes forming image of the sun.

**4.** A → air, glass, polythene
B → tracing paper, plastic scale, muddy water
C → cardboard, brick, earth

| | | | |
|---|---|---|---|
| **5.** (b) | **6.** (a) | **7.** (c) | **8.** (c) |
| **9.** (C) | **10.** (c) | **11.** (b) | **12.** (b) |
| **13.** (d) | **14.** (d) | **15.** (a) | |

Chapter

# 5 ELECTRICITY

## KEY FACTS

1. **Electricity** is a form of energy that can be easily changed to other forms. That is one of the reasons why it has so many uses.
2. Electrical energy is produced by electric current. A current is produced when electric charges flow.
3. **Electric charge**. Electric charges can be positive or negative. All matter is made up of very small particles called ***atoms***. Atoms are made of even smaller particles called ***protons***, ***neutrons***, and ***electrons***. Protons are positively charged and electrons are negatively charged particles, while neutrons are not charged, i.e., they are neutral.

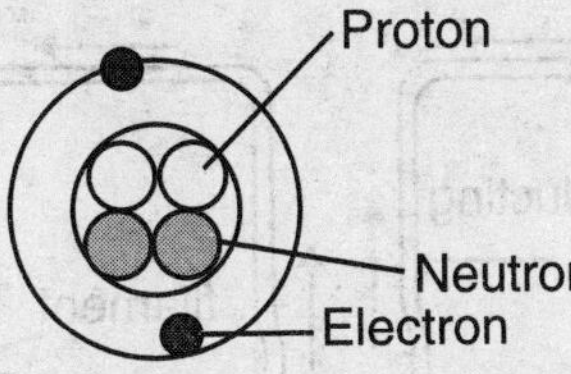


*Protons and neutrons make up the centre of the atom. Electrons are found outside the nucleus.*

4. Charge is a physical property. Charged objects exert a force – a push or a pull - on other charged objects. **The law of electric charges** states that ***like charges repel, or push away, and opposite charges attract.***

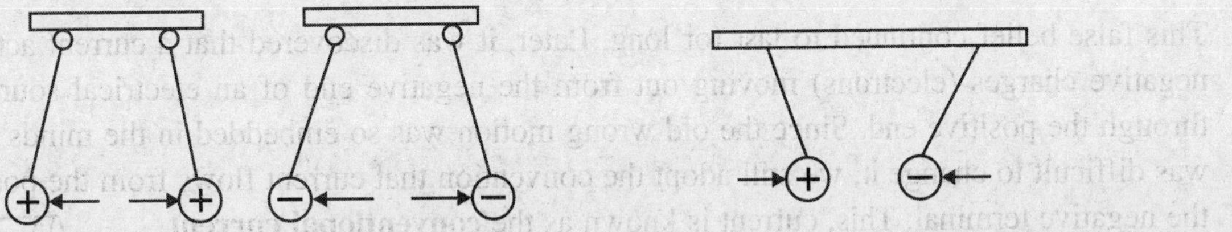

*Objects that have the same charge repel each other.* *Objects that have opposite charges attract each other.*

5. **Atoms have equal number of protons and electrons**. Because an atoms' positive and negative charges cancel each other, atoms do not have a charge. So, how does an object made up of atoms be charged. An object becomes positively charged when it loses electrons. An object becomes negatively charged when it gains electrons. This can be done by various methods like friction etc.

> It may be noted that an atom can lose or gain only the electrons, which are negatively charged. There is no loss or gain of protons.

6. A current is produced when electric charges flow.
   Electrons flow from a body having excess electrons to a body that has a deficit of electrons.
   The SI unit for electric current is the **ampere (A)**.

Electric currents are measured by instruments called **ammeters**.

7. An **electrical circuit** is the path along which electrical charges move. It consists of a source of electrical energy (such as a battery), connecting wires and one or more electrical components such as switch, bulb etc. The electric current flows out from the source, round the circuit and back, unless there is a break in the circuit.

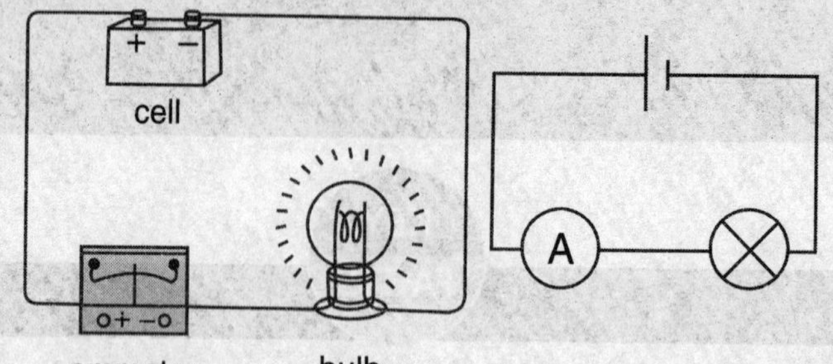


8. **Direction of the current**. When the flow of current in metal wires began to be studied in the past, it was believed that a current was a flow of positive charges moving out from the positive end of a source of electrical energy and returning to the negative end.

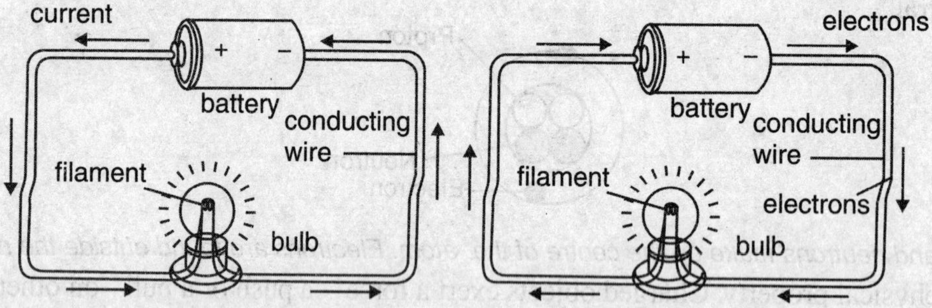


*Flow of conventional current* *Actual flow of electrons*

This false belief continued to last for long. Later, it was discovered that a current actually consists of negative charges (electrons) moving out from the negative end of an electrical source and returning through the positive end. Since the old wrong motion was so embedded in the minds of people that, it was difficult to change it, we still adopt the convention that current flows from the positive terminal to the negative terminal. This, current is known as the **conventional current**.

9. **Open and Closed Circuits**
   (1) When the electrons flow from the negative end of the source (battery) through the wire to the bulb and back to the battery, thus completing the circuit, then it is called a **closed circuit**.
   (2) A circuit where there is a break such that electric current cannot flow through it is called an **open circuit**.

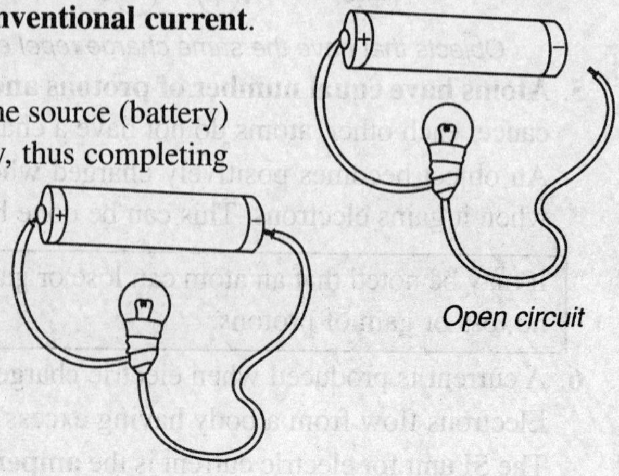

*Open circuit*

*Closed circuit*

Using symbols, closed and open circuits can be shown as under:

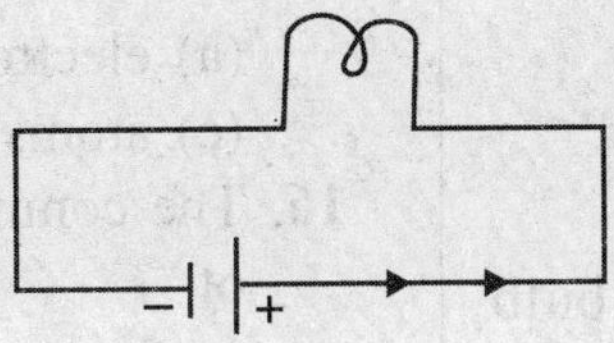

Closed circuit

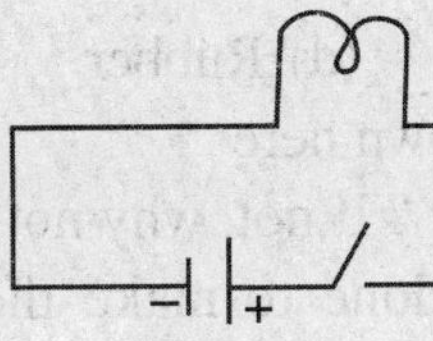

Open circuit

**10. Conductors and Insulators**

(1) In atoms of some materials, the electrical force is not strong enough to hold the electrons to the nucleus, and hence the electrons are able to move about in the material. Material having electrons which move around freely in it allows the electric charge to pass easily through it and is called a **good conductor** of electricity or simply *conductor*.
Metals such as iron, steel, copper, and aluminium are good conductors. Most wire used to carry a charge is made of copper, which conducts very well.

(2) Materials that do not have electrons that are free to move through them do not allow an electric charge to pass easily through them. They are called **insulators**. Plastic and rubber are good insulators. Many types of electric wire are covered with plastic which insulates as well.

(3) Pure water is not a conductor of electricity. Salty water is a conductor of electricity. A substance that conducts electricity in solution is called an **electrolyte**. Here, salt is an electrolyte when it forms solution with water. Note that dry salt is not a conductor of electricity.

## Question Bank–5

**1. Fill in the blank in each of the following:**

(i) The path along which electricity travels is called a ________.

(ii) A device that is used to break or make an electric circuit is called ________ .

(iii) An electric cell has ________ terminals.

(iv) An electric circuit in which there is a gap is called an ________ circuit.

(v) An electric circuit in which there is no gap is called a ________ circuit.

(vi) Air is not a conductor of electricity. It is an ________.

(vii) A substance that conducts electricity in a solution is called an ________ .

(viii) Electricity is the flow of electric ________.

(ix) An atom contains protons, neutrons and ________.

(x) A ________ is a material that allows an electric charge to pass through it easily.

(xi) An ________ is a material that does not easily allow a charge to pass through it.

(xii) Electric charge is a ________ of matter.

(xiii) Electrons have a ________ charge.

(xiv) Protons have a ________ charge.

(xv) An electric current is a ________ of electric charges.

**2. Choose the letter of the best answer:**

An electric charge is a

(a) kind of liquid

(b) type of matter

(c) kind of chemical reaction

(d) force acting at a distance

**3.** Which of the following statements is true?

(a) Electricity can be created

(b) Electricity flows in a circuit with gaps

(c) Electricity is the flow of negative charge

(d) All of the above

**4.** What happens to a circuit when the switch is off?

(a) The circuit is complete

(b) There is a gap in the circuit

(c) Electricity flows continuously

(d) Electricity flows discontinuously

**5.** Two like charges

(a) repel each other

(b) first repel then attract

(c) first attract then repel

(d) attract each other

6. Which of the following is not a bad conductor?
   (a) Mica (b) Iron
   (c) Wood (d) Rubber
7. Look at the circuit shown here
   (a) Will the bulb glow ? If not, why not?
   (b) What should be done to make the bulb glow?

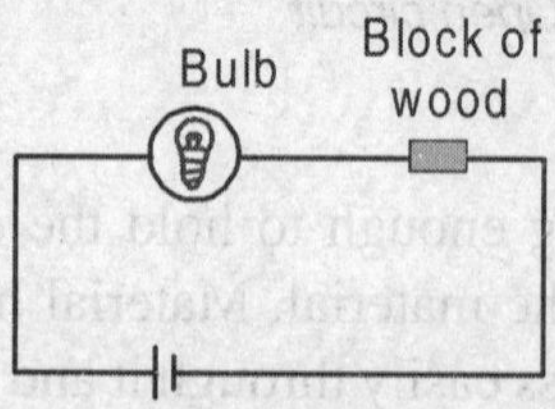


8. Draw wires to connet the bulb to the battery so that the bulb glows with maximum brightness.

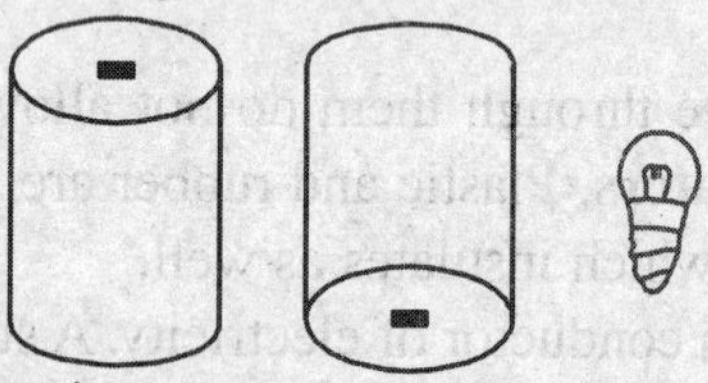

9. The electric current is produced due to
   (a) flow of electrons
   (b) flow of protons
   (c) flow of positive and negative charges
   (d) vibration of atoms
10. Answer '**True or False**' for the following statements.
   (a) Electric current can flow through metals
   (b) Instead of metal wires, a jute string can be used to make a circuit.
   (c) Electric current can pass through a sheet of thermocol
11. The two places on a battery to which the circuit wires can be attached are called.
   (a) switch (b) filament
   (c) terminals (d) insulators
12. Using the 'conduction tester' on an object it was found that the bulb begins to glow. Is that object a conductor or an insulator? Explain.
13. The conventional direction of an electric current is:
   (a) from excess electrons to deficient electrons
   (b) from deficient electrons to excess electrons
   (c) from high temperature to low temperature
   (d) current has no direction.
14. When few electrons are removed from a neutral body, the body is charged
   (a) positive (b) negative
   (c) neutral (d) none of the above
15. What constitutes current in a metal wire?
   (a) electrons (b) protons
   (c) atoms (d) molecules
16. The commonly used safety fuse wire is made of
   (a) lead
   (b) copper
   (c) nickel
   (d) an alloy of tin and lead
17. Priyanka connected two different pieces of material and enclosed them in a box with three external connections A, B, C as shown. She then asked Rekha to guess how the two pieces of material were connected.

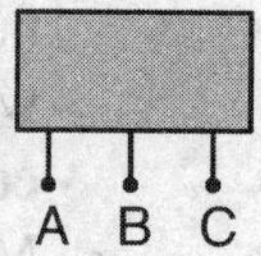


Rekha conducted the following experiments with the black box.

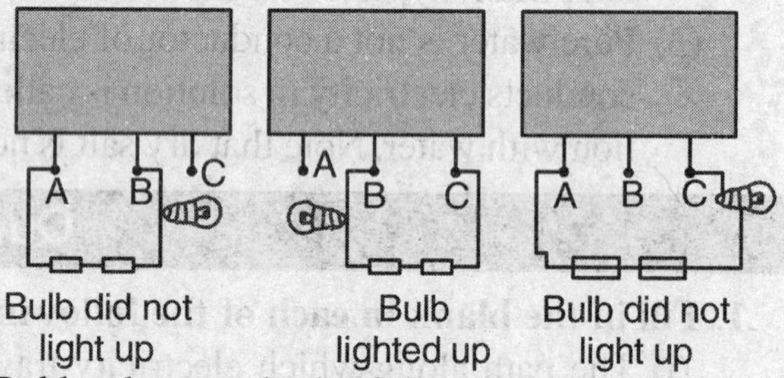


Rekha then made the following four gueses of which only one is correct. Which one is that ?

(a)
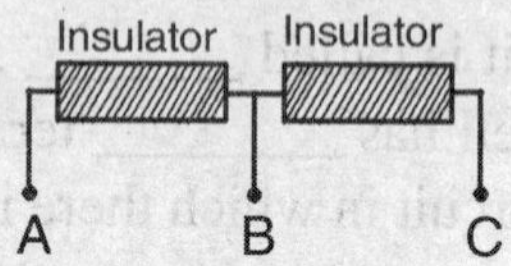


(b)
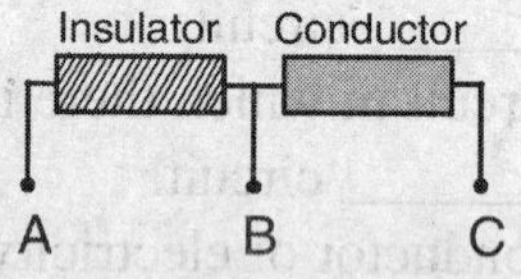


(c)
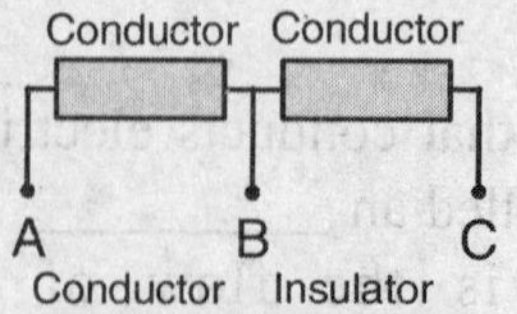


(d)
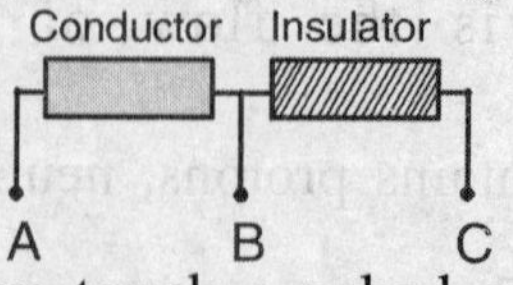


18. If we touch a naked current carrying wire, we get a shock. This happens because our body is a:
   (a) insulator of electricity
   (b) source of electricity
   (c) conductor of electricity
   (d) both (a) and (b)

**19.** Which of the following is the best conductor of electricity?
(a) ordinary water
(b) sea water (c) boiled water
(d) distilled water (e) rain water

**20.** The metal used in storage batteries is
(a) iron (b) copper
(c) lead (d) tin

**21.** The filament of an electric bulb is made of
(a) iron (b) michrome
(c) tungsten (d) graphite

**22.** The handle of every repairing tool is covered by a certain insulating material so that the user may not get an electric shock. Which of the following materials cannot be used to cover the handle ?
(a) Wood (b) Plastic
(c) Glass (d) Tin

**23.** Overhead cables need not be insulated because
(a) air is a bad conductor of electricity
(b) air is a good conductor of electricity
(c) bare wires conduct electricity better than insulated wires
(d) none of these

**24.** Look at the two electric circuits using materials A and B as shown below.

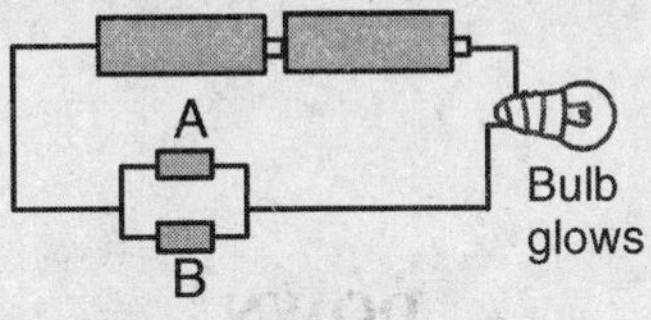


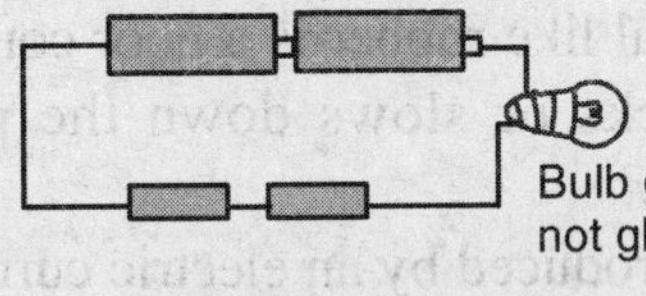


What do you conclude from the above ?
(a) A is a conductor and B is an insulator ?
(b) A is an insulator and B is a conductor
(c) Either A or B is an insulator
(d) Both A and B are conductors

**25.** Safety wire, used in electrical circuits, is made of a material having
(a) low melting point
(b) high melting point
(c) high resistance

**26.** The best conductor of electricity is
(a) Aluminium (b) Iron
(c) Copper (d) Silver

**27.** Which of the following statements is true?
(a) Lightning never hits trees
(b) Riding a bicycle during lightning is safe
(c) Going for a swim during a lightning strike is safe
(d) Tall objects are more prone to lightning strikes

**28.** Which one of the following cells is rechargeable?
(a) Daniell cell (b) Leclanche cell
(c) Lead cell (d) Volta cell

**29.** Tungsten is used for the manufacture of the filament of an electric bulb because
(a) it is a good conductor
(b) it is economical
(c) it is malleable
(d) it has a very high melting point

**30.** Conversion of chemical energy into electrical energy occurs in
(a) dynamos (b) electric heaters
(c) battery (d) atomic bombs

**31.** A current is passed through a vertical spring from whose lower end a weight is hanging. What will happen to the weight?
(a) The weight shall go up
(b) The weight shall go down
(c) The position of the weight will remain the same
(d) The weight shall oscillate

**32.** Lightning conductors are made of
(a) Iron (b) Copper
(c) Steel (d) Chromium

**33.** The acid used in lead storage cells is
(a) phosphoric acid (b) hydrochloric acid
(c) nitric acid (d) sulphuric acid

**34.** Consider the following statements:
(1) An isolated electric charge exists but an isolated magnetic monopole does not exist.
(2) Electric lines of force are not closed, but magnetic lines of force are closed.
Which of the statements given above is/are correct?
(a) 1 only (b) 2 only
(c) Both 1 and 2 (d) Neither 1 nor 2

**35.** Which particles can be added to the nucleus of an atom without changing its chemical properties?
(a) Electrons (b) Protons
(c) Neutrons (d) None of the above

**36. Solve the following crossword with the help of the given clues :**

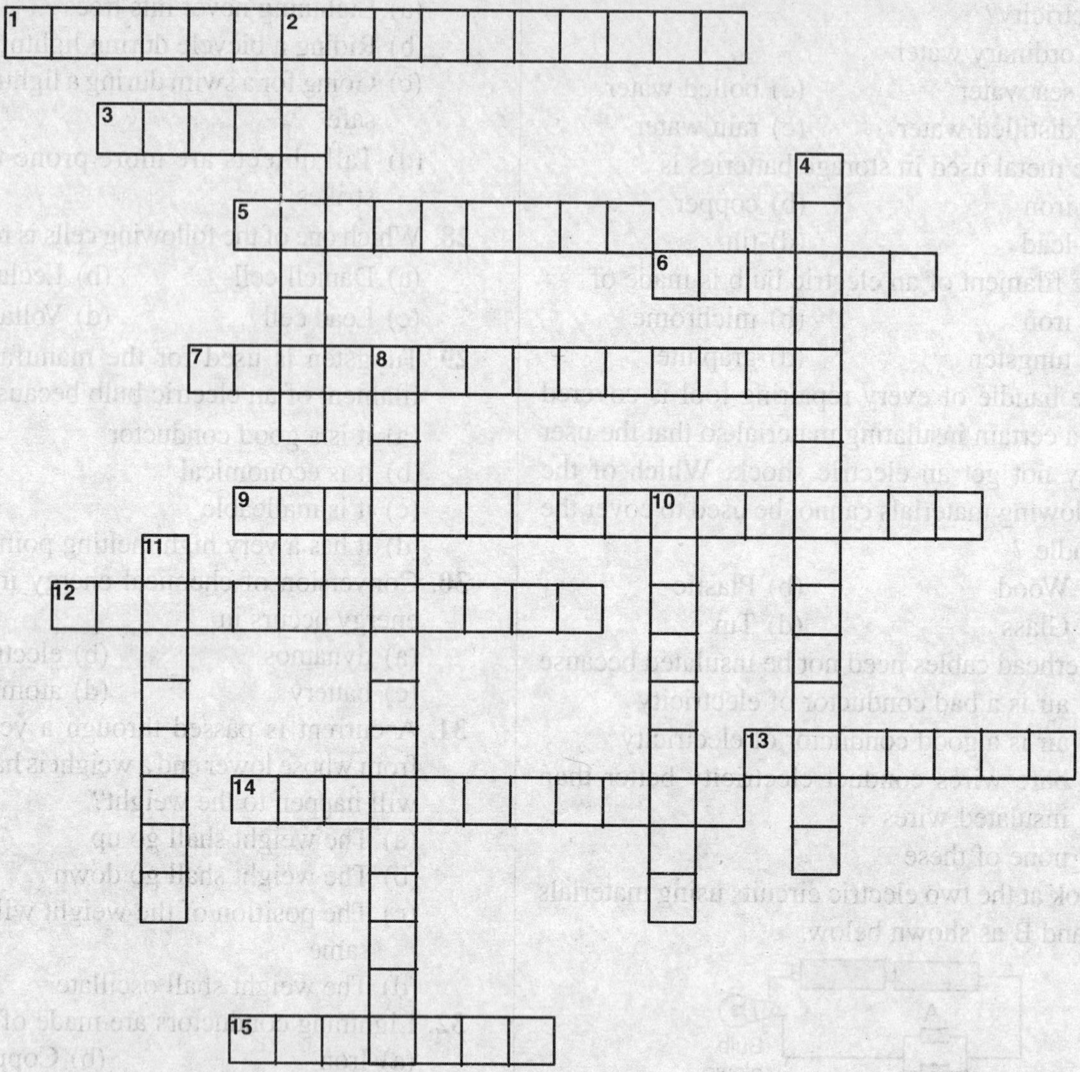


### ACROSS

**1.** The flow of electricity through a material.
**3.** Smallest particles that make up the basic structure of a chemical element.
**5.** Elementary particles in an atom which carry a negative electric charge and revolve around the nucleus of the atom.
**6.** The SI unit for electric current
**7.** An electric circuit through which current can flow in an uninterrupted path.
**9.** The path along which electrical charges move.
**12.** An electric circuit through which current cannot flow because of a break in the path.
**13.** Particles in the nucleus of an atom. They have a positive electric charge.
**14.** A substance that can conduct electricity in a solution.
**15.** An instrument that measures the strength of an electric current in units called amperes.

### DOWN

**2.** A material like rubber is a poor conductor and that blocks or slows down the passage of electricity.
**4.** Energy produced by an electric current.
**8.** Charge relating to electricity that is made up of positive and negative charge.
**10.** Materials (like copper) or objects that allows electric charge to flow easily through it.
**11.** One of the particles having no electric charge that make up an atom.

## Answers

1. (i) circuit (ii) switch (iii) two
(iv) open (v) closed (vi) insulator
(vii) electrolyte (viii) charge (ix) electrons
(x) conductor (xi) insulator (xii) property
(xiii) negative (xiv) positve (xv) flow

2. (d) 3. (c) 4. (b) 5. (a) 6. (b)

7. **(a)** No, because wood being an insulator would not allow the current to flow through it.
**(b)** The block of wood should be replaced by a metallic wire.

8.

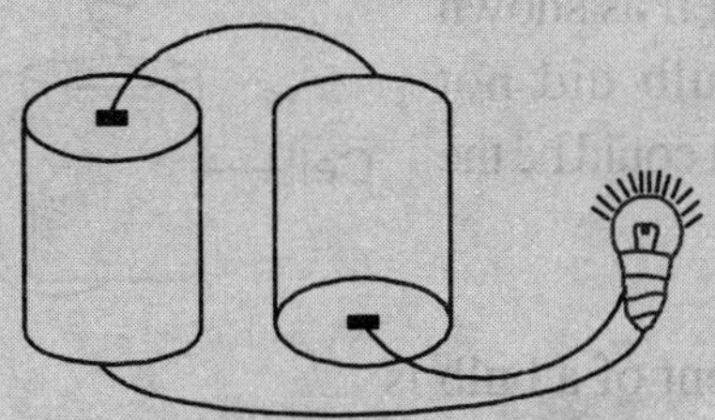

9. (a)

10. (a) True (b) False (c) False

11. (c)

12. That object is a conductor because electricity can pass through only a conductor and not through an insulator. The bulb would glow only if the object is a conductor.

13. (a) 14. (a) 15. (a) 16. (b)
17. (b) 18. (c) 19. (b) 20. (c)
21. (c) 22. (d) 23. (a) 24. (c)
25. (a) 26. (d) 27. (d) 28. (c)
29. (d) 30. (c) 31. (d) 32. (b)
33. (d) 34. (c) 35. (c)

36.

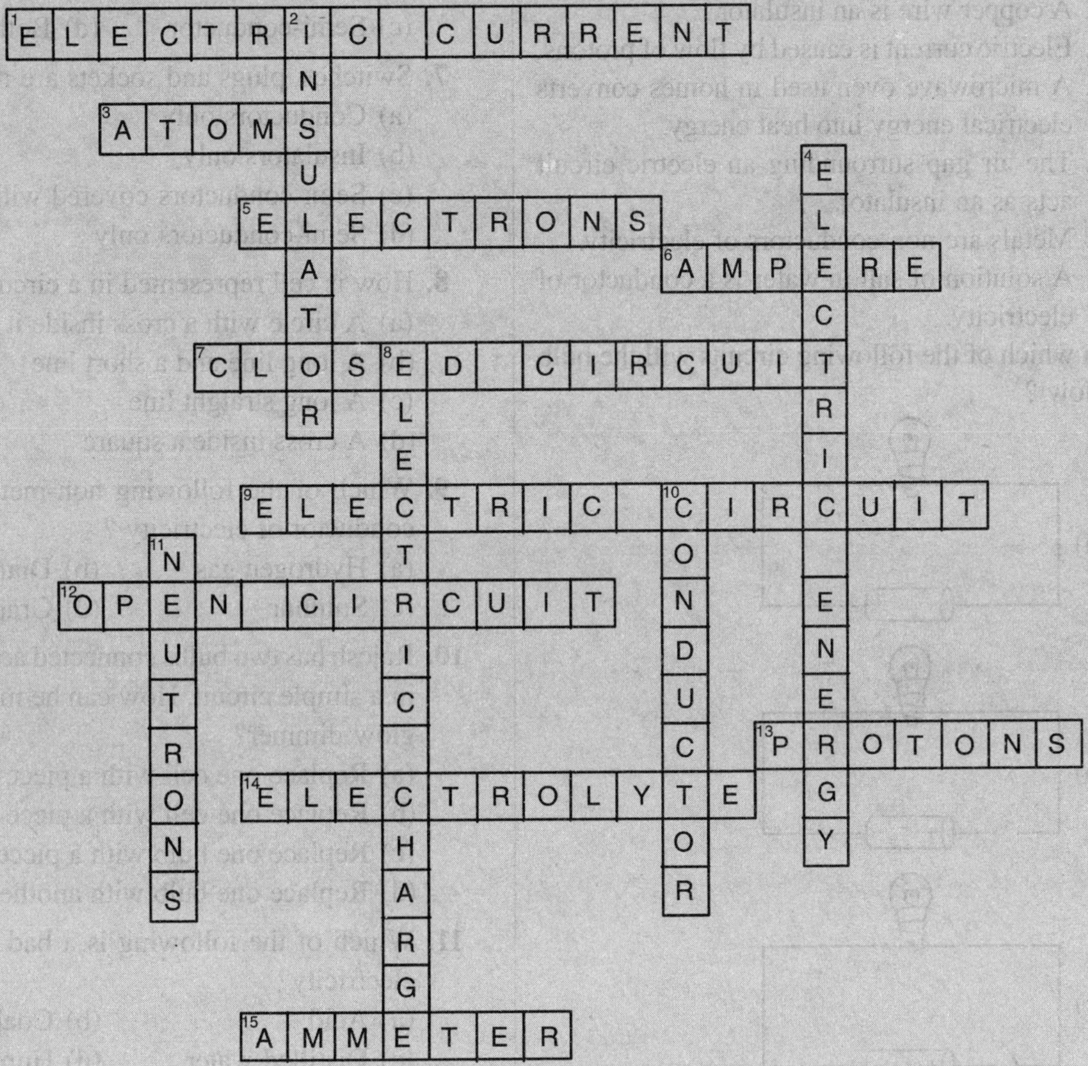

## Self Assessment Sheet–5

**1. Match correctly**

| *Column I* | *Column II* |
|---|---|
| 1. Insulator | a. cell |
| 2. Used for making filament of a bulb | b. switch |
| 3. Conductor | c. plastic |
| 4. Converts chemical energy into electrical energy | d. graphite |
| 5. Device used to break or complete a circuit | e. battery |
| 6. Group of two or more cells | f. Tungsten |

**2. Answer True or False**

1. Human body is a bad conductor of electricity.
2. A cell always has two terminals.
3. A copper wire is an insulator.
4. Electric current is caused by flow of protons.
5. A microwave oven used in homes converts electrical energy into heat energy.
6. The air gap surrounding an electric circuit acts as an insulator.
7. Metals are non-conductors of electricity.
8. A solution of salt in water is a conductor of electricity.

**3.** In which of the following circuits will the bulb glow ?

(a)

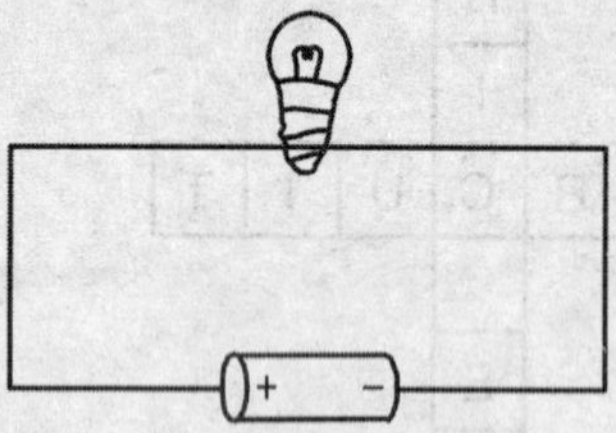


(b)

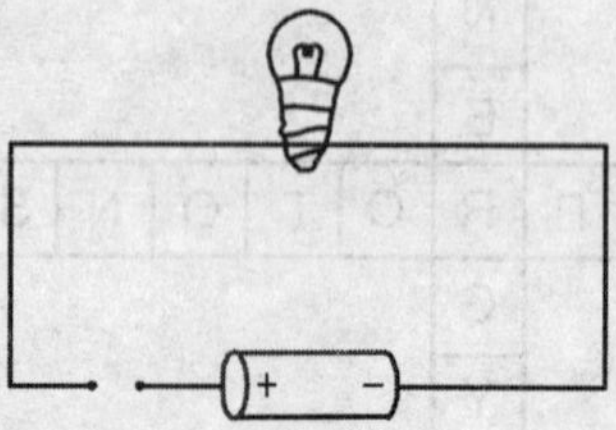


(c)

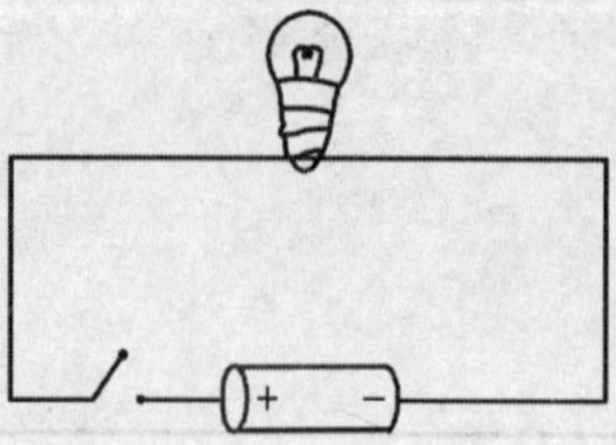


**4.** A student connected the terminals P and Q by a piece of thread. Will the bulb glow?

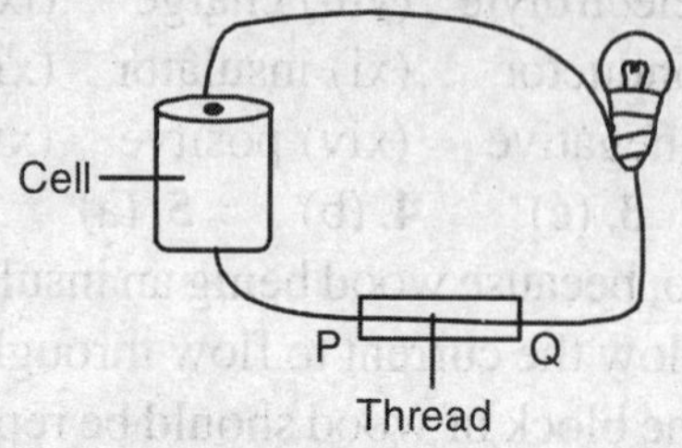


**5.** Irshad connected a bulb to a cell as shown but the bulb did not glow. What could be the reason?

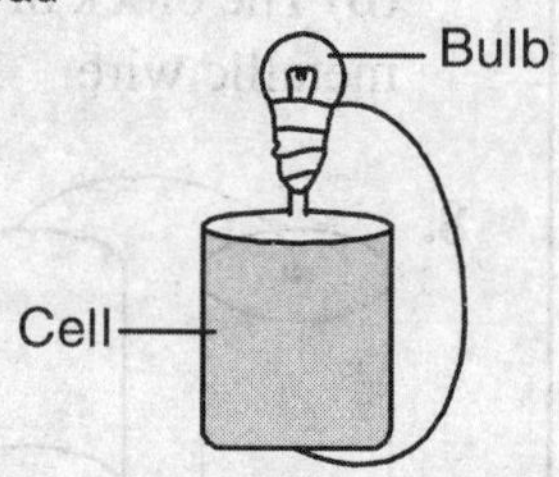


**6.** The filament of a bulb is
(a) Conductor (b) Insulator
(c) Semi-conductor (d) Both (a) and (b)

**7.** Switches, plugs and sockets are made of
(a) Conductors only
(b) Insulators only
(c) Semi-conductors covered with insulators
(d) Semi-conductors only

**8.** How is cell represented in a circuit diagram
(a) A circle with a cross inside it
(b) A long line and a short line
(c) A long straight line
(d) A cross inside a square

**9.** Which of the following non-metals is a good conductor of electricity ?
(a) Hydrogen gas (b) Diamond
(c) Sulphur (d) Graphite

**10.** Rajesh has two bulbs connected across two cells in a simple circuit. How can he make the bulbs glow dimmer?
(a) Replace one cell with a piece of chalk
(b) Replace one cell with a piece of wire
(c) Replace one bulb with a piece of wire
(d) Replace one bulb with another cell

**11.** Which of the following is a bad conductor of electricity?
(a) Acid (b) Coal
(c) Distilled water (d) Human body

12. The metallic cap projecting out from the top of a cell is
(a) switch (b) positive terminal
(c) negative terminal (d) spring

13. All liquids are non-conductor of electricity execpt
(a) mobil oil (b) coconut oil
(c) glycerine (d) mercury

14. Pure water is a non-conductor of electricity but can be made conductor by adding to it :
(a) sugar (b) drops of oil
(c) salt (d) pieces of wood

15. Which of the following is the correct way of setting a circuit ?

(a)
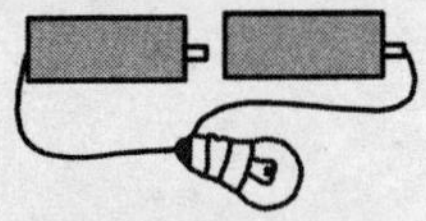

(b)
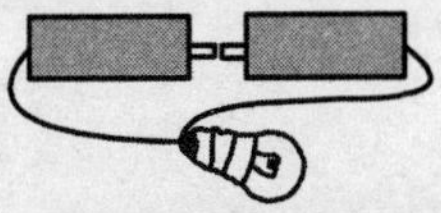

(c)
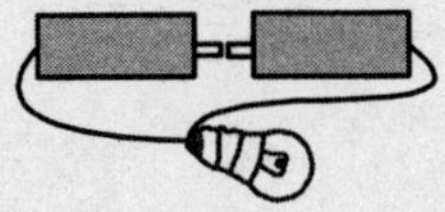

(d)
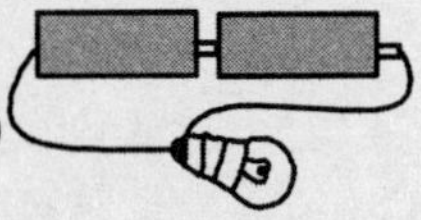

## Answers

1. (i) –c, (2) –f (3) –d, (4) –a, (5) –b, (6) –e

2.

| 1 | 2 | 3 | 4 | 5 | 6 | 7 | 8 |
|---|---|---|---|---|---|---|---|
| F | T | F | F | T | T | F | T |

3. (a)

4. No, the bulb will not glow because thread being an insulator breaks the circuit.

5. (i) The bulb could be fused, *i.e.*, its filament was broken.
(ii) The cell was dead, *i.e.*, its chemical was used up.

6. (a) 7. (c) 8. (b) 9. (d) 10. (b)
11. (c) 12. (b) 13. (d) 14. (c) 15. (d)

# Chapter 6

# MAGNETISM

## KEY FACTS

1. The materials which are attracted towards a magnet are called **magnetic**.
2. Any object that has the property of attracting magnetic metals like iron, cobalt and nickel is called a **magnet**.
3. Magnets can have many shapes. They may be cylindrical bar magnets, rectangular bar magnets, horse-shoe or U-shaped magnets, or ring magnets.
4. A material which is attracted by a magnet is called a **magnetic material**.
5. The regions of a magnet where the attraction of magnet is the strongest are called poles of the magnet. They are near its free ends and are always different.
6. One pole of the magnet is called the **north pole** and the other is called the **south pole**.
7. A freely suspended magnet always points in North-South direction.
8. The **compass** is an instrument used to find directions at a place.
9. **Like magnetic poles repel each other and unlike magnetic poles attract each other.**
10. **A magnet loses its magnetism if it is dropped from a height, hammered or heated.**
11. **The earth acts as a big magnet.** It pulls every object around it. The imaginary magnet inside the earth is believed to be lying in the direction of geographic North and geographic South, in a little inclined position. This means that the magnetic pole of the Earth, that is close to the geographic North pole is a magnetic *South* pole and Earth magnetic North pole is close to the Earth geographic South pole.
12. A soft iron bar placed in line with a magnet acts like a magnet itself. This is **induced magnetism**.
13. **Magnetic poles cannot be isolated. They can exist only in pairs.**
14. The area surrounding a magnet up to which the magnetic force acts is called a **magnetic field**. The magnetic field is maximum at the poles of a bar magnet.

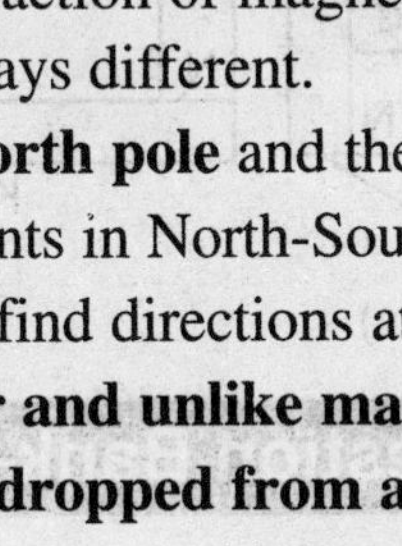

**15.** **Temporary magnets** are easy to magnetise but lose their magnetisation easily.
**Permanent magnets** are difficult to magnetise but retain their magnetic properties for a long time.

**16.** All matter is made of atoms. The atoms contain negatively charged particles called electrons. As an electron moves around, it makes, or induces, a magnetic field. The atom will then have a north and a south pole. In most materials, such as copper and aluminium, the magnetic fields of the individual atoms cancel each other out. Therefore, these materials are not magnetic.
But in materials such as iron, nickel and cobalt, groups of atoms are in tiny areas called *domains*. The north and south poles of the atoms in a domain line up and make a strong field. Domains are like tiny magnets of different sizes within an object.

**17.** There are different kinds of magnets. Some magnets are made of iron, nickel, cobalt, or mixtures of these metals. Magnets made with these metals have strong magnetic properties and are called ***ferromagnets***. Magnets made by passing an electric current through a substance are called ***electromagnets***. They remain as a magnet as long as the current passes through them.

**18.** **Magnetic keepers.** Magnets need to be stored properly to prevent loss of their magnetism over a period of time. Bar magnets are kept in pairs with unlike poles facing each other. They are separated by a non-magnetic material like wood. Two soft iron bars (called keepers) are placed across their ends as shown in the figure. A horse shoe magnet is stored by placing a piece of soft iron bar across the poles.

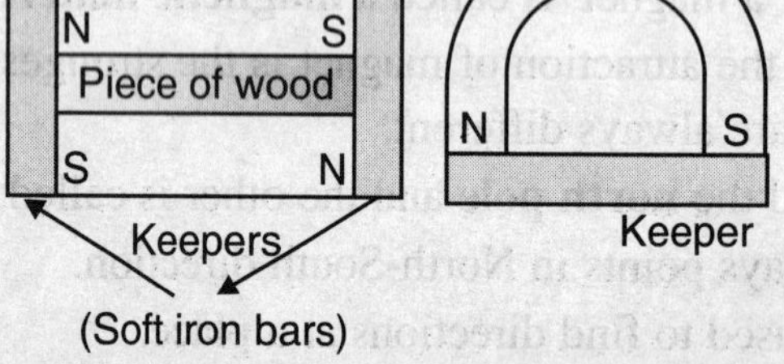


## Question Bank–6

**1. State whether the following statements are true or false:**

(i) It is possible to have a magnet like a cylindrical magnet having only one pole.
(ii) Some non-magnetic materials can be converted into magnets.
(iii) Like poles of magnets repel each other.
(iv) Maximum iron filings stick in the middle of the bar magnet when it is brought near them.
(v) Bar magnets always point towards North - South direction.
(vi) A compass can be used to find East - West direction at any place.
(vii) Rubber is a magnetic material.
(viii) Magnetic property is destroyed when a magnet is heated strongly.
(ix) A compass helps in determining directions.
(x) If a magnet is cut into two halves, the north pole will get separated from the south pole.

**2.** For each of the cases in the figure below, identify whether the magnets will attract or repel one another.

(a) 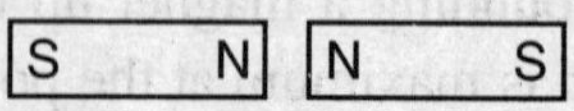


(b) 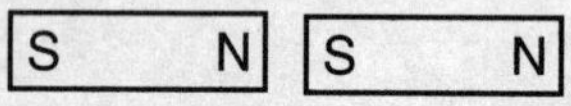


(c) 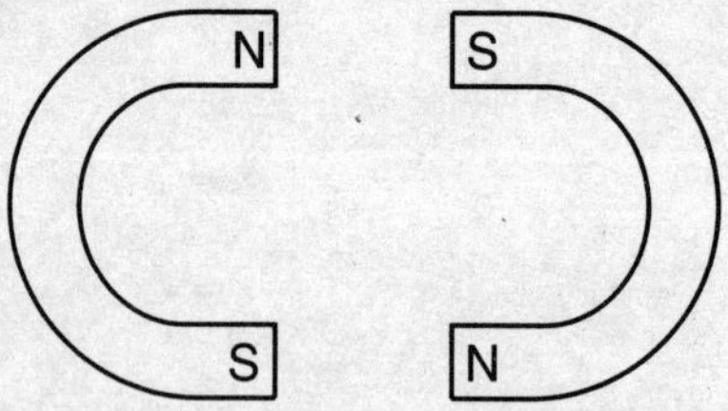


**3.** When you break a bar magnet in half, how many poles does each piece have?

4. If two magnets push each other away, what can you conclude about their poles?

5. Which magnetic pole is closest to the geographic north pole?

6. Name three properties of magnets.

7. What metal is used to make ferromagnets?

(a) iron (b) cobalt

(c) nickel (d) all of the above.

8. A freely suspended magnet always lies in

(a) North-South direction

(b) North-East direction

(c) East-West direction

(d) South-East direction

9. The magnetic field is maximum at

(a) middle part of the magnet

(b) all places in the magnet

(c) poles of the magnet

(d) none of these.

10. Magnetic poles always exist as

(a) dipole (b) monopole

(c) no-pole (d) none of these

11. Which substance among the following is a magnetic substance

(a) copper (b) iron

(c) silver (d) aluminium

**12. Match correctly:**

| *Column I* | *Column II* |
|---|---|
| (i) A magnet is dropped from a height on a hard rock | (a) magnetism is induced in the soft iron bar |
| (ii) An electric current is passed into a coil of copper wire wound around a soft iron bar | (b) magnetism is lost |
| (iii) A soft iron bar is placed in line with a magnet | (c) the iron bar is magnetised |
| (iv) A strong magnet is stroked on an iron bar from one end to the other a number of times. | |

13. A bar magnet has no markings to indicate its poles. How would you find out near which end is its north pole located?

14. When any magnet is cut into four equal parts and then they are again joined by quick fix then new magnet will behave as

(a) four bar magnets

(b) four ordinary rods

(c) one ordinary bar magnet

(d) one ordinary iron rod

15. An iron rod is considered as a magnet if the north pole of any other magnet

(a) repels its both the ends

(b) attracts its both the ends

(c) neither attracts nor repels any of its ends

(d) attracts its one end and repels its other end

16. Is the magnetic field of Earth stronger near the middle of Earth (in Mexico) or at the bottom of Earth (in Antarctica)? Explain your answer.

17. Why are some iron objects magnetic and others not magnetic?

18. Where is the magnetism maximum in case of horse-shoe magnets?

(a) In both the poles

(b) In south pole

(c) In north pole

(d) In between the two poles

19. Which of the following devices is used to find the direction by soldiers?

(a) Magnetic pencil (b) Magnetic compass

(c) Bar magnet (d) Electromagnet

20. How are temporary magnets different from permanent magnets?

**21**. In which part of a bar magnet, the magnetisation is zero?

(a) At the centre (b) At the poles
(c) Both the ends (d) None of these

**22. Consider the following statements**:

As one moves from one place to another, the magnetic field of the earth will vary

(1) in magnitude
(2) in direction
(3) linearly with height
(4) linearly with the temperature of the place of these statements

(a) 1 and 2 are correct
(b) 3 alone is correct
(c) 1, 2, 3 and 4 are correct
(d) 3 and 4 are correct

**23**. Consider the following statements

(1) The magnetic pole in the northern hemisphere is the north magnetic pole.
(2) At all points on a magnet, an iron bar gets attracted.

Which of the statements given above is/are correct?

(a) 1 only (b) 2 only
(c) Both 1 and 2 (d) Neither 1 nor 2

**24**. If a magnetic needle is freely suspended at the geographic north pole

(a) The needle will remain vertical with its N-pole downward
(b) The needle will remain almost vertical with its N-pole downward
(c) The needle will remain vertical with its S-pole downward
(d) The needle will remain almost vertical with its S-pole downward

**25**. What is the number of neutral points for a bar magnet with its north pole pointing geographical north?

(a) Zero (b) One
(c) Two (d) Infinite

**26**. A bar magnet with poles N and S marked is freely suspended. Then the end marked N would point towards

(a) the South magnetic pole of the earth
(b) the South geographic pole of the earth
(c) the North magnetic pole of the earth
(d) the North geographic pole of the earth

**27**. Choose the wrong statement:

(a) Single magnetic pole can exist
(b) Magnetic poles are always of equal strength
(c) Like poles repel each other
(d) None of these

**28**. Magnetism in materials is due to

(a) electrons at rest
(b) motion of electrons
(c) protons at rest
(d) all neutrons at rest

**29**. Consider the following statements:

(1) At the centre of the bar magnet, magnetism is zero.
(2) Induced magnetism is always lost completely on removing the inducing field.
(3) A pocket watch can be shielded from magnetic effects if it is kept inside an iron casting.

Which of the statements given above are correct?

(a) 1 and 2 (b) 1 and 3
(c) 2 and 3 (d) 1, 2 and 3

**30**. Prepare a concept map by using the following words :

*Materials, Magnetic, Non-magnetic, Magneıs-Permanent, Temporary, Electro-magnets, Properties, Uses, Loss of magnetism.*

**31. Solve the following crossword with the help of the given clues.**

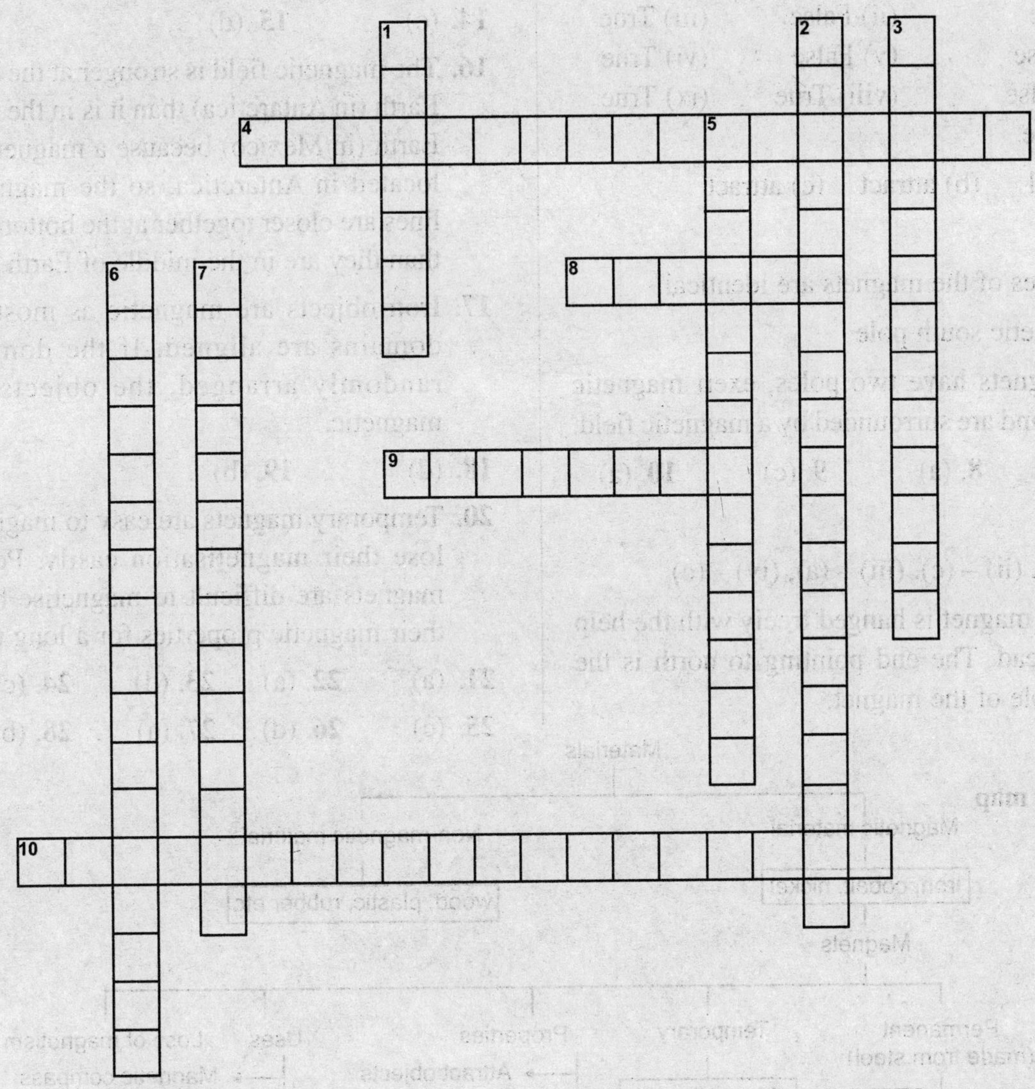


**ACROSS**

**4.** Materials that retain their magnetism permanently.

**8.** A piece of metal or other solid that has the property of attracting iron or steel.

**9.** Having the properties of a magnet.

**10.** Region of magnets near South pole where magnetism is the strongest.

**DOWN**

**1.** A device with a needle used for determining direction.

**2.** Region of magnet near North pole where magnetism is the strongest.

**3.** A soft iron bar that behaves like a magnet when current flows through the coil surrounding the iron bar.

**5.** The area around a magnet in which objects are affected by the force of the magnet.

**6.** Materials that retain their magnetism only for a short period of time.

**7.** The two opposite areas of a magnets where the magnetism is strongest.

## Answers

**1.** (i) False (ii) False (iii) True
(iv) False (v) False (vi) True
(vii) False (viii) True (ix) True
(x) False

**2.** (a) repel (b) attract (c) attract

**3.** two

**4.** The poles of the magnets are identical.

**5.** A magnetic south pole

**6.** All magnets have two poles, exert magnetic forces, and are surrounded by a magnetic field.

**7.** (d) **8.** (a) **9.** (c) **10.** (a)

**11.** (b)

**12.** (i) – (b), (ii) – (c), (iii) – (a), (iv) – (c)

**13.** The bar magnet is hanged freely with the help of a thread. The end pointing to north is the north pole of the magnet.

**14.** (c) **15.** (d)

**16.** The magnetic field is stronger at the bottom of Earth (in Antarctica) than it is in the middle of Earth (in Mexico) because a magnetic pole is located in Antarctica, so the magnetic field lines are closer together at the bottom of Earth than they are in the middle of Earth.

**17.** Iron objects are magnetic as most of their domains are aligned. If the domains are randomly arranged, the objects are not magnetic.

**18.** (d) **19.** (b)

**20.** Temporary magnets are easy to magnetise but lose their magnetisation easily. Permanent magnets are difficult to magnetise but retain their magnetic properties for a long time.

**21.** (a) **22.** (a) **23.** (d) **24.** (c)

**25.** (c) **26.** (d) **27.** (a) **28.** (b)

**29.** (b)

**30. Concept map**

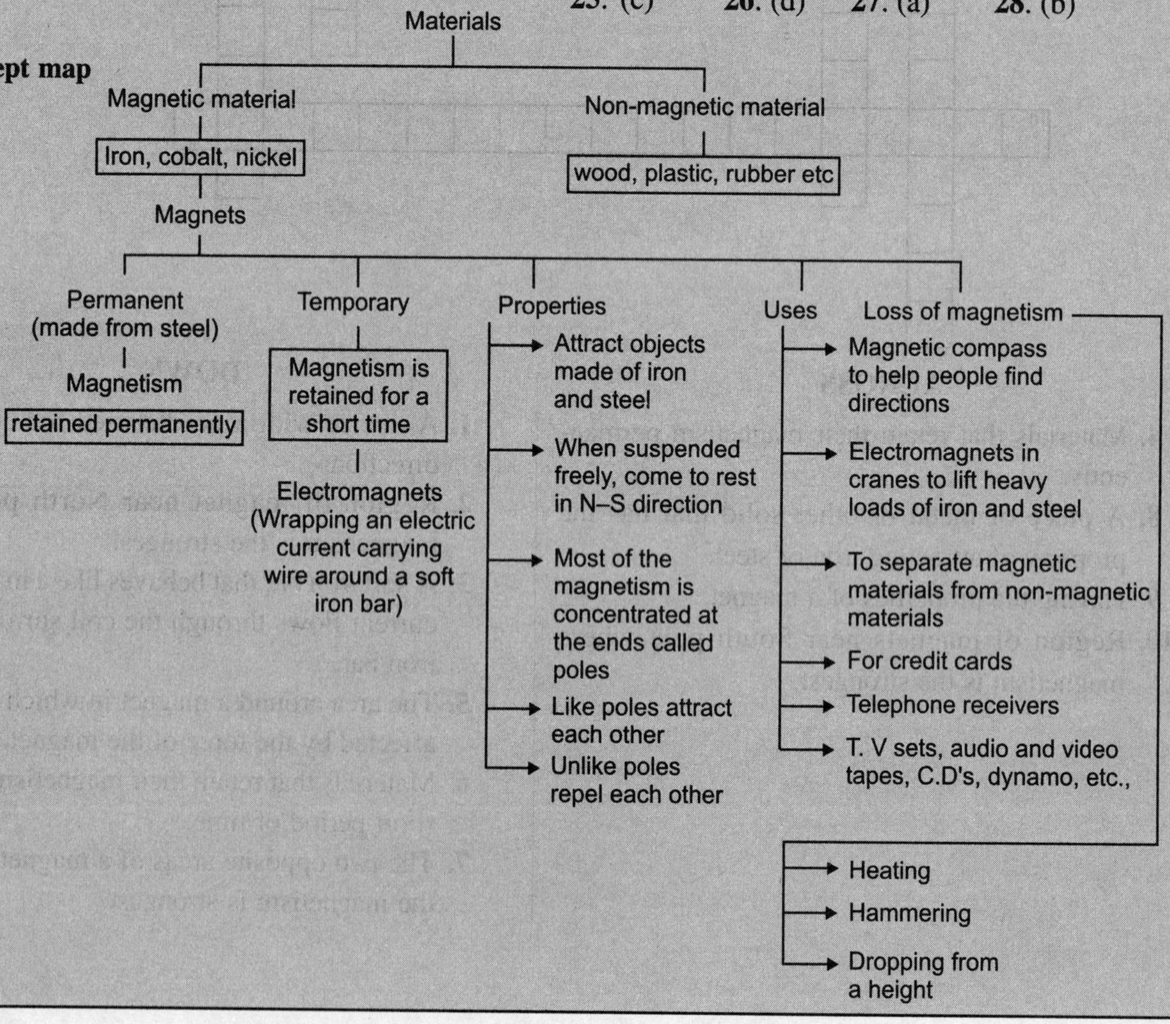

**31.**

## Self Assessment Sheet-6

**1. Choose the correct word and fill in the blanks.**

1. Glass is a ______ (magnetic/non-magnetic) material.
2. The magnetism is ______ (least/strongest) in the middle of a magnet.
3. Like poles of magnets ______ (repel/attract).
4. A freely suspended bar magnet comes to rest in the ______ (East-West, North-South) direction.
5. When materials like iron are magnetised a ______ (temporary/permanent) magnet is formed.

**2.** State two ways that can cause a magnet to lose its properties.

**3. State whether the following statements are true or false.**

(1) A copper coin is a magnetic material.

(2) A magnet can attract an iron nail from a distance.

(3) The force of pull/attraction of a magnet is maximum in the middle.

(4) The earth magnetic north pole is near the geographic south pole.

(5) Iron, nickel and cobalt are the only three magnetic metals.

(6) Heating increases the magnetic power of a magnet.

(7) Induced magnetism is temporary.

**4. Answer the following questions.**

(1) What is piece of iron inside a wire coil called.

(2) What happens when you turn off the current?

(3) How can you increase the strength of an electromagnet?

(4) You can make an electromagnet by wrapping a coil around the nail and magnetise it by flowing a current through the coil. Can you use an aluminium nail instead of an iron nail? Why?

**5.** A small magnet is suspended by a silk thread from a rigid support such that the magnet can freely swing. How will it rest ? Draw a diagram to show it.

**6.** Four all pins are hanged from the north pole of a bar magnet. Mark the magnetic poles induced in these pins. Can these pins hang even if the bar magnet is removed?

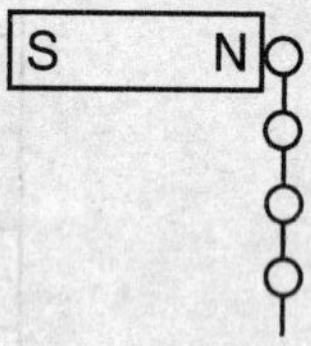


**7.** Two long needles are attached to the poles of a horse shoe magnet. Show on a diagram the position occupied by the needles and name the phenomenon which comes into play.

**8.** Show the poles in the broken parts of the magnets when it is divided into two halves

(i) along the axis XX′.

(ii) along the axis YY′.

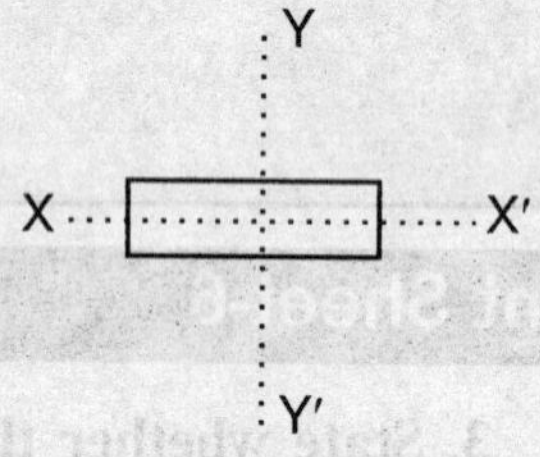


**9.** Metal bars are brought near each pole of a compass needle in turn. Complete the following table :

| | Nature of bar | Action on compass needle | |
|---|---|---|---|
| | | North pole | South pole |
| (i) | Non-magnetic like glass | No action | No action |
| (ii) | ---------- | Attracted | Attracted |
| (iii) | North pole of a bar magnet | ------------ | ------------ |
| (iv) | ---------- | Attracted | Repelled |

**10.** A horse-shoe magnet, when not in use, is kept with a metal piece A, that is held to the north and south poles.

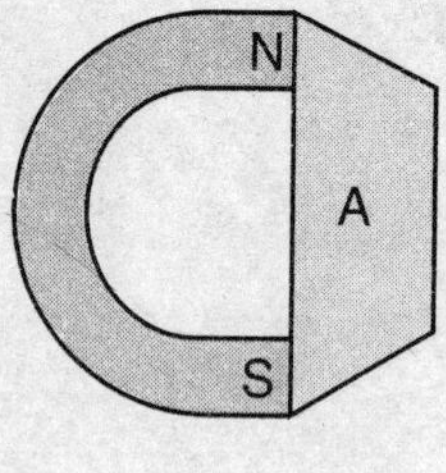


(i) What is the metal piece A called ?

(ii) What is A made of ?

(iii) What is the function of the piece A ?

**11.** The middle region of a bar magnet is

(a) a north pole

(b) a north seeking pole

(c) unmagnetised

(d) magnetised

**12.** A magnet pulls some substances towards it. This means :

(a) The magnet exerts some force

(b) The magnet can do some work

(c) Both (a) and (b)

(d) None of the above

**13.** A small magnet Y is placed near a heavy magnet X as shown in the figure. How will the magnets move ?

(a) magnet Y will move away

(b) magnet Y will turn clockwise

(c) magnet Y will turn anti-clockwise

(d) Both the magnets will turn clockwise

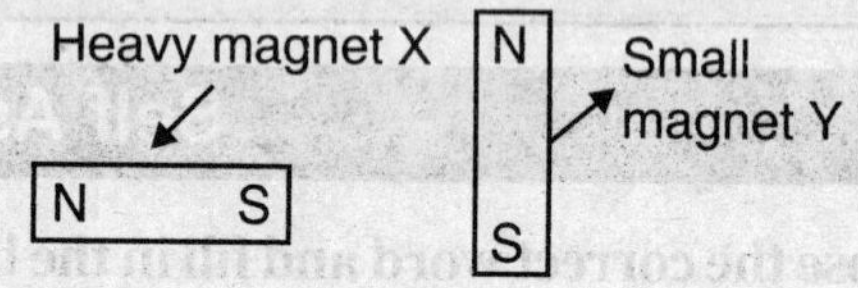


**14.** When one end of an iron bar is placed near a compass :

(a) it is always the north pole of the compass that points towards it

(b) it is always the south pole of the compass that points towards it

(c) any one of the poles of the compass would point towards it

(d) the compasss will remain unaffected by the iron bar.

**15.** Two ring magnets P and Q each with a hole in the centre are dropped one over the other on a plastic rod taking care that like poles of the magnets face each other. One magnet say Q comes to the bottom of the rod and it would appear that the other magnet say P floats above the first gap leaving a gap in between. It is so because

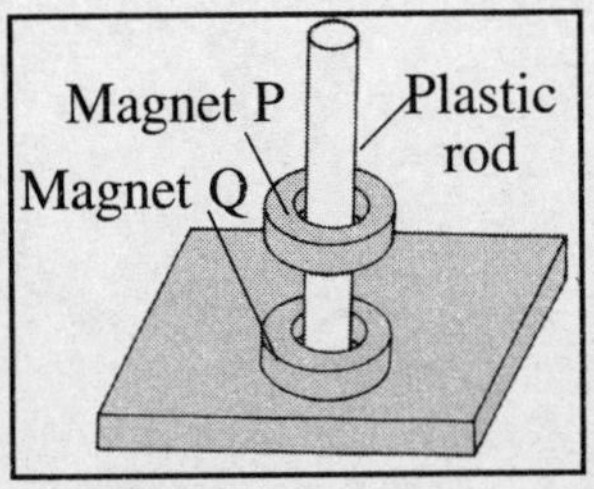


(a) magnet P is lighter than magnet Q
(b) magnet Q is more powerful than magnet P.
(c) Similar poles of the two magnets repel each other when placed face to face. The repulsion causes the rise of one of the magnets.
(d) both (b) and (c)

## Answers

**1.** (a) Non-magnetic (b) least
(c) repel (d) North-South
(e) permanent

**2.** (i) Heating (ii) Hammering

**3.** (1) False (2) True (3) False
(4) True (5) True (6) False
(7) True

**4.** (i) Electromagnet (ii) The magnetic field disappears. The metal acts as a magnet as long as the coil carries a current. (iii) By increasing the number of coils or the current (iv) No, an aluminium foil would not form a magnet, because aluminium is not a magnetic material.

**5.** It will rest in geographic north-south direction with north pole towards, the geographic north, making some angle with the horizontal as shown in the figure.

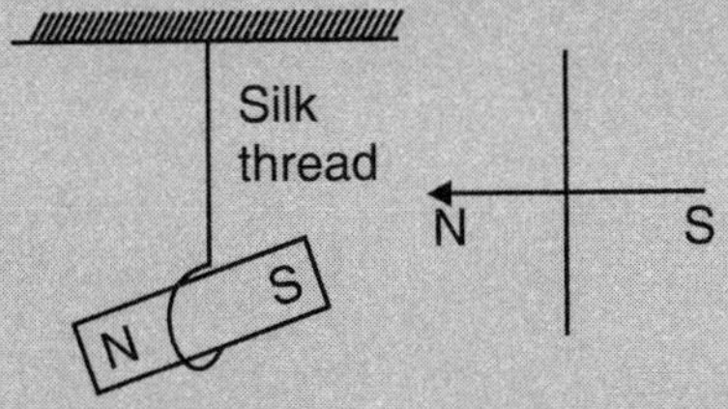


**6.** Magnetism is induced in all pins. No, the pins will get separated when the bar magnet is removed.

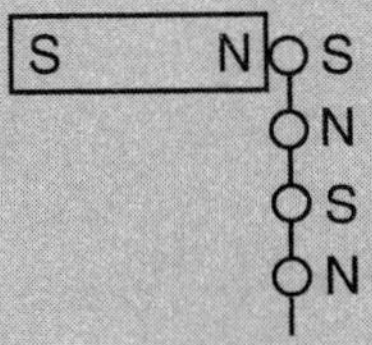


**7.** The needles will get attached in the positions as shown. The upper ends touching the poles of the magnet would acquire opposite polarities to those of the magnet. The lower ends of the needles will get attracted towards each other or touch each other since they have acquired opposite polarities. The phenomenon is called magnetic induction.

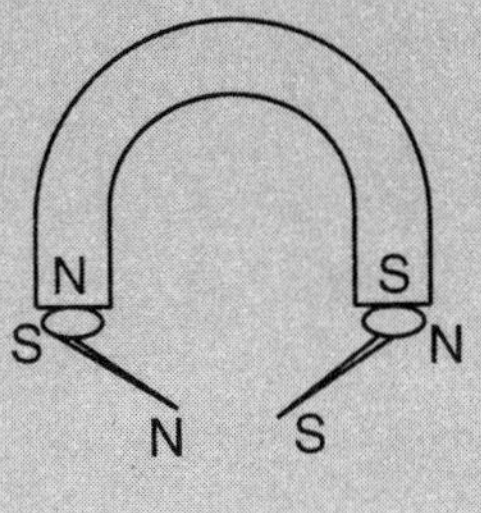


**8.** **(i)**

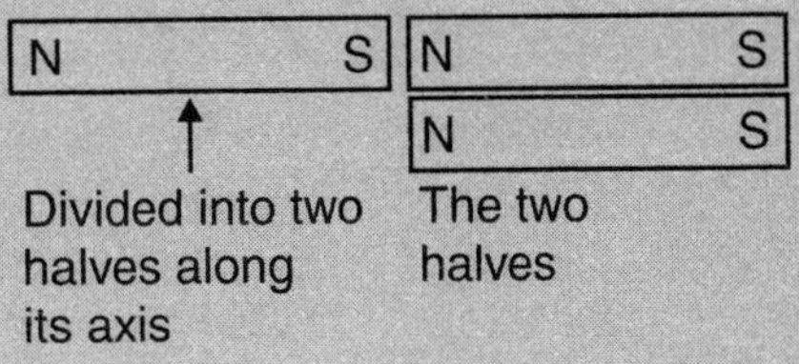


**(ii)**

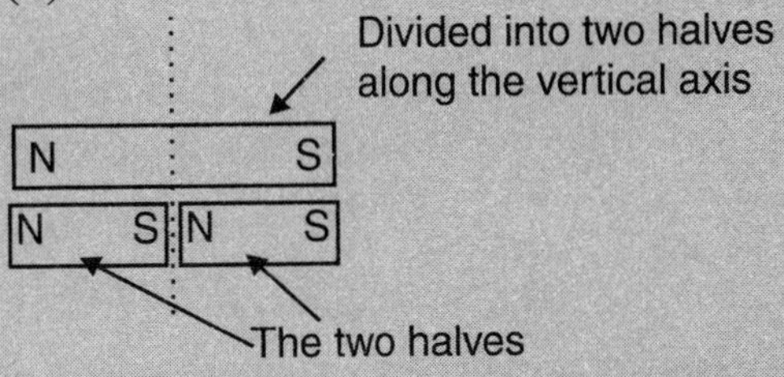


| **9.** | **Nature of bar** | **Action on compass needle** | |
|---|---|---|---|
| | | **North pole** | **South pole** |
| **(i)** | Non-magnetic | No action | No action |
| **(ii)** | Iron | Attracted | Attracted |
| **(iii)** | North pole of a bar magnet | Repelled | Attracted |
| **(iv)** | South pole of a bar magnet | Attracted | Repelled |

**10.** (i) Magnetic keeper
(ii) soft iron
(iii) as preserver of magnetism in the magnet.

**11.** (c) **12.** (c) **13.** (c)

**14.** (c) **15.** (d)

# TOPICS IN CHEMISTRY

- *Sorting Materials into Groups*
- *Separation of Substances*
- *Changes Around Us*
- *Water and Air*

Chapter 7

# SORTING MATERIALS INTO GROUPS

## KEY FACTS

1. Anything that has mass and occupies space is called **matter**.
2. The matter of which an object is made is called **material**. There are many kinds of materials. Wood, nylon, fibre, paper, metals, water, cement, etc., are all materials.
3. Objects or materials are often recognised from their **physical properties**.
   A **physical property** is a characteristic of matter that can be observed without causing any change to the matter.
4. Materials are of two types: **natural materials** and **man-made materials**.
   (i) Wood, Coal, Petroleum, Water, Wool, Silk, Gold, Rocks etc., are natural materials.
   (ii) Iron, Cement, Glass, Plastics, Sugar, Paints, etc., are man-made materials.
5. An object may be made of the same material, different materials or a combination of materials.
   **For example**, a chair may be made of wood, or plastic or metal or concrete.
   Concrete is made by combining cement, gravel, sand and water.
6. **Properties of materials.** Colour, size and shape are some common physical properties of matter which are clearly visible. Other physical properties of matter include ***density*** (heaviness or lightness with respect to water), ***strength, appearance*** (shiny or dull), ***hardness*** or ***softness, flexibility, solubility or insolubility in water, transparency, translucency or opaqueness, electrical conductivity, thermal conductivity, boiling point, melting point,*** etc.
7. The strength and hardness of a material is its ability to support heavy load and withstand scratches and wear. Steel and concrete are such materials. Hard materials may be **brittle**, or **malleable** and **ductile.** Soft material may be elastic.
   (i) **Brittle**–A material like chalk that is likely to break, snap, crack or become powdered.
   (ii) **Elasticity**–A material which is capable of returning to the original shape and size after having been stretched, squeezed or bent is said to be elastic. Rubber bands are elastic.
   (iii) **Plasticity**–The materials like wet clay and dough which change their shape under the influence of force but do not return to the original shape when the force is withdrawn are called **plasticines**.
   (iv) **Malleability**–Metals like gold, silver and copper which are capable of being hammered and beaten into sheets are known as **malleable**.
   (v) **Ductility**–Metals such as gold when heated can be easily drawn out into thin wires. They are called **ductile** materials.

**8. Description of physical properties which form the basis of classification.**

(i) **Dense** – Materials like stone which are heavier than water sink in it. They are denser than water. Materials like cork which float on water are lighter than water. They are less dense than water.

(ii) The **flexibility** of a material is its ability to bend without breaking.

(iii) Materials like gold, silver, diamond, aluminium, copper, etc., which have shiny appearance are said to have **lustre** and are called lustrous.

Materials like wood, paper, plastic, etc., have dull appearance, *i.e.*, they do not have shine.

(iv)(a) Materials like iron, steel, stone, brick, diamond, etc., which cannot be easily compressed, cut, bent (moulded) or scratched are called **hard materials**. Diamond is the hardest known material.

(b) Materials like sponge, rubber, clay, feathers, candle wax, etc., which can be easily compressed, cut, bent or scratched are called **soft materials**.

(v) **Solubility in water**.

(a) Substances like sugar and salt are **soluble** in water and form a transparent solution. Such substances just disappear in water.

(b) Substances like sand, a piece of wood, hair (fibres) which remain as such when put in water are said to be **insoluble** in water. Sand settles down at the bottom of water while wood floats on its surface.

(c) The substance which gets dissolved in a liquid is called a **solute**. The liquid in which the substance is dissolved is called a **solvent** and the resultant product of dissolving solute in a solvent is called a **solution**. Water is a solvent.

(d) Liquids like milk and water which mix completely are called **miscible** liquids. Liquids like oil and water which do not mix completely are said to be **immiscible**.

(vi) **Transparency, Translucency and Opaqueness**

(a) Materials like glass, air, water etc., through which we can see clearly are called **transparent** materials.

(b) Materials like ground-glass, tracing paper, etc., through which we cannot see clearly are called **translucent** materials.

(c) Materials like cardboard, metals, brick, book, etc., through which we cannot see at all are called **opaque** materials.

(vii) **Combustibility.** Combustible materials catch fire easily, *e.g.*, kerosene, petrol, wood.

(viii) The **thermal conductivity** of a material is a measure of how readily heat flows through it. Materials like metals that allow heat to pass through them easily are called **thermal conductors**. Materials such as wood, plastics and gases that do not allow heat to flow through them easily are called **thermal insulators**.

(ix) **The electrical conductivity** of a material is a measure of how readily electricity flows through the material. Those that do not allow electricity to flow through them easily are called **insulators**. Metals are good electrical conductors while non-metals with the exception of carbon (graphite) are good electrical insulators.

**9. Diffusion**. The spreading out of a gas is called diffusion. It takes place in a haphazard and random manner. It is due to diffusion that we can smell the perfumes.

**10.** Different materials have different sets of properties which make them useful for particular jobs. Five main classes of materials are listed below.

(i) **Non-metals**

**Non-metals**

- **Plastics** made from petroleum, e.g., PVC, bakelite
- **Fibres** can be found naturally or made artificially, e.g., cotton, silk, wool, rayon, jute
- **Glass** made from heating soda, sand and limestone
- **Ceramics** made from clay, e.g., pottery

(ii) **Metals**, *e.g.*, iron, gold, aluminium, titanium, lead, copper and zinc found naturally in its pure form or in the form of ore as in iron.

## Question Bank-7

**Fill in the blanks with the correct word(s) :**

**1. A metal**

(i) has a ________ melting point.

(ii) has ________ surfaces.

(iii) can be beaten into different shapes and bent without ________.

(iv) is a ________ conductor of heat.

(v) is a ________ conductor of electricity.

**2. Glass**

(i) has ________ definite melting point.

(ii) does ________ corrode or rot.

(iii) is a ________ and ________supercooled liquid.

(iv) can be moulded into different shapes at ________ temperatures.

**3. Plastics**

(i) are made from ________.

(ii) does ________ corrode or rot and has a ________ density.

(iii) can be moulded into different shapes ________ solidifying.

(iv) is a ________ conductor of heat.

(v) is a ________ conductor of electricity.

**4. Ceramics**

(i) is made from ________.

(ii) has a ________ melting point and ________ corrode or rot.

(iii) is a ________ and ________ solid.

(iv) can be moulded into different ________ at high temperatures.

(v) is a ________ conductor of heat and electricity.

**5. Fibre**

(i) can be spun into ________ and woven into ________.

(ii) is an ________ and ________ solid.

(iii) can be coloured according to the colours of the absorbed ________.

(iv) is a ________ conductor of heat and electricity.

**6.** Match the objects given below with the materials from which they could be made.

| Objects | Materials |
|---|---|
| Book | Glass |
| Tumbler | Wood |
| Chair | Paper |
| Toy | Leather |
| Shoes | Plastics |

**7. Match each word with its definition.**

| | |
|---|---|
| (i) flexibility | (a) it determines whether a substance will sink or float in water (= mass per unit volume) |
| (ii) malleability | (b) a substance which catches fire easily. |
| (iii) hardness | (c) the ability of a material to bend without breaking |
| (iv) strength | (d) the ability of a material to return to its original shape and size after the pressure put on it is withdrawn |
| (v) density | (e) the ability of a material to support heavy loads without breaking or tearing |
| (vi) electrical conductivity | (f) capable of being beaten or hammered into sheets |

(vii) elasticity (g) the ability of a material to withstand scratches and wear

(viii) combustible (h) a measure of how readily electricity flows through a material.

**8. Find the odd one out from the following.**

(a) Chair, Bed, Table, Baby, Cupboard
(b) Rose, Jasmine, Boat, Marigold, Lotus
(c) Aluminium, Iron, Copper, Silver, Sand
(d) Sugar, Salt, Sand, Milk, Milk powder
(e) Nail, Utensils, Magnet, Beaker, Coin
(f) Wood, Stone, Iron, Cotton, Diamond

**9.** Will the following items float or sink if they are put on water, oil or kerosene?
Plastic ball, Feather, Matchstick, Balloon, Wood, Hair, Thermocole, Boat

**10.** Which of the following is a solution?

(a) soil (b) rocks
(c) lemonade (d) noodle soup

**11.** An example of a solution is

(a) sea water (b) vegetable soup
(c) sand and water (d) concrete

**12.** Which of the following is a good conductor of heat?

(a) silver (b) paper
(c) wood (d) plastic

**13.** The property of being drawn into wire is called

(a) malleability (b) conductivity
(c) combustibility (d) ductility

**14.** Which of the following is a bad conductor of electricity?

(a) copper (b) steel
(c) impure water (d) air

**15.** Overhead cables need not be insulated because

(a) air is a bad conductor of electricity
(b) air is a good conductor of electricity
(c) bare wires conduct electricity better than insulated wires
(d) it is not economical to insulate them

**16. State whether the following statements are true or false.**

(1) Sulphur and graphite dissolve in water
(2) Wood and plastic are insulators
(3) Air is transparent
(4) A notebook has lustre while eraser does not
(5) Chalk dissolves in water
(6) Vinegar dissolves in water
(7) A piece of wood floats on water
(8) Oil mixes with water
(9) Sand settles down in water
(10) Sugar does not dissolve in water

**17. Fill in the following blanks with suitable words.**

(1) Mustard oil and water are ________ liquids.
(2) Water and milk are ________ liquids.
(3) A piece of glass is transparent but a piece of cardboard is ________.
(4) Pure water is a ________ conductor of electricity.
(5) An oily white paper in relation to light is ________.
(6) The smell of a perfume reaches us by the process of ________.
(7) Kitchen utensils have wooden handles since wood is a ________ of heat.

**18.** What material should be used to make the lid opener?

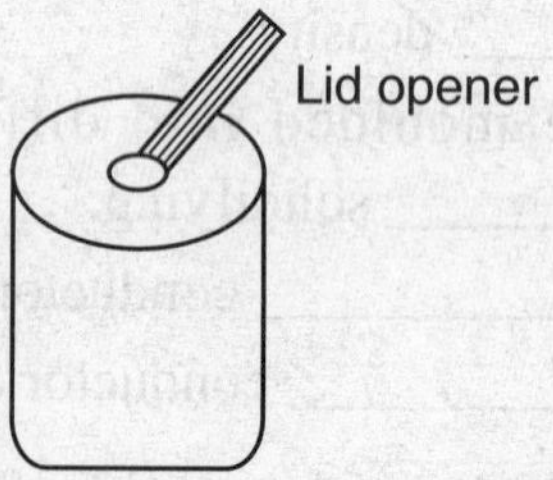


(a) Rubber (b) Metal
(c) Fabric (d) Plastic

**19.** Study the following statements.

**1.** A paper weight is hard. It therefore sinks in water.

**2.** Plastic is soft. It can be scratched easily.

Now, tell which of the following are valid statements?

(a) Statement 1 is correct for all hard substances
(b) All soft substances can be scratched easily.
(c) All soft objects will float on water.
(d) Both statements are wrong.

**20.** The beakers shown contain equal amounts of water and another material. Which list shows the solubility of the different materials in the beakers, from most to least soluble?

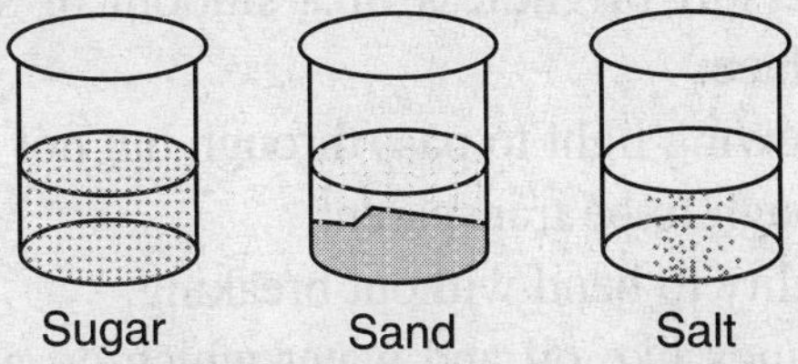


(a) salt, sand, sugar (b) salt, sugar, sand
(c) sand, salt, sugar (d) sugar, salt, sand

**21.** What is common between the following pairs of substances
(a) copper and silver (b) wood and air
(c) water and mercury (d) LPG and petrol

**22.** Which property of gas helps us in detecting leakage of cooking gas (LPG).

**23. Solve the following crossword with the help of the given clues :**

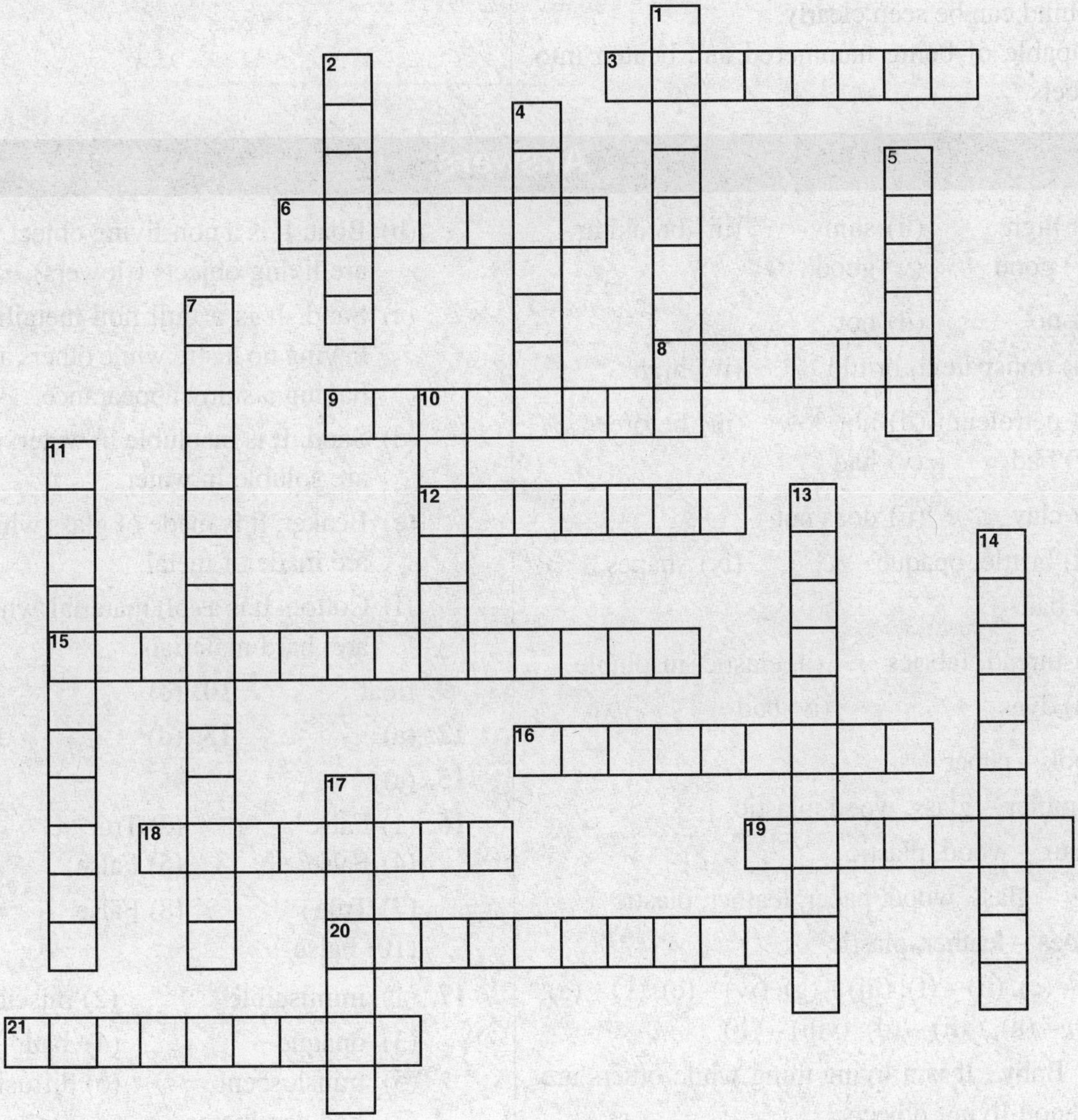

### ACROSS

3. Liquids such as water and alcohol which mix completely in any proportion.
6. Easily drawn out into a fine strand or wire.
8. Not letting light pass through.
9. A liquid in which something is dissolved.
12. Capable of being dissolved.
15. Materials which can be easily compressed, cut, bent or scratched.
16. Substances such as sand which do not dissolve in water.
18. The substance or things from which something is or can be made.
19. Capable of going back to its original length or shape after being stretched or squeezed.
20. Allowing light to pass through so that objects behind can be seen clearly.
21. Capable of being hammered and beaten into sheets.

### DOWN

1. The spreading out of a gas.
2. Substance which gets dissolved in a liquid.
4. A substance in which another substance is dissolved, forming a solution.
5. Something that occupies space, has mass, and can exist as solid, liquid, or gas.
7. Materials which cannot be easily scratched.
10. The soft brightness of a smooth or shining surface.
11. Allowing light to pass through but not clearly enough to be transparent.
13. Ability to bend without breaking.
14. Liquids like oil and water which do not mix completely.
17. Hard but likely to break, snap, or crack.

## Answers

**1.** (i) high (ii) shiny (iii) breaking
(iv) good (v) good

**2.** (i) no (ii) not
(iii) transparent, brittle (iv) high

**3.** (i) petroleum (ii) not, low (iii) before
(iv) bad (v) bad

**4.** (i) clay (ii) does not
(iii) brittle, opaque (iv) shapes
(v) bad

**5.** (i) thread, fabrics (ii) elastic, insoluble
(iii) dyes (iv) bad

**6.** Book – paper
Tumbler – glass, wood, plastic
Chair – wood, plastic
Toy – glass, wood, paper, leather, plastic
Shoes – leather, plastic

**7.** (i) – (c), (ii) – (f), (iii) – (g), (iv) – (e), (v) – (a), (vi) – (h), (vii) – (d), (viii) – (b)

**8.** (a) Baby. It is a living thing while others are non-living objects.
(b) Boat. It is a non-living object while others are living objects (flowers).
(c) Sand. It is a dull non-metallic substance having no lustre while others are all metals having a shiny appearance.
(d) Sand. It is insoluble in water while others are soluble in water.
(e) Beaker. It is made of glass while all others are made of metal.
(f) Cotton. It is a soft material while all others are hard materials.

**9.** float **10.** (c) **11.** (a)
**12.** (a) **13.** (d) **14.** (d)
**15.** (a)

**16.** (1) False (2) True (3) True
(4) False (5) False (6) True
(7) True (8) False (9) True
(10) False

**17.** (1) immiscible (2) miscible
(3) opaque (4) bad
(5) transluscent (6) diffusion
(7) bad conductor

**18.** Metal, since the opener should be hard and strong.

**19.** (b) **20.** (d)

**21.** (1) They are metals (2) insulators (3) liquids (4) combustible

**22.** Diffusion

**23.**

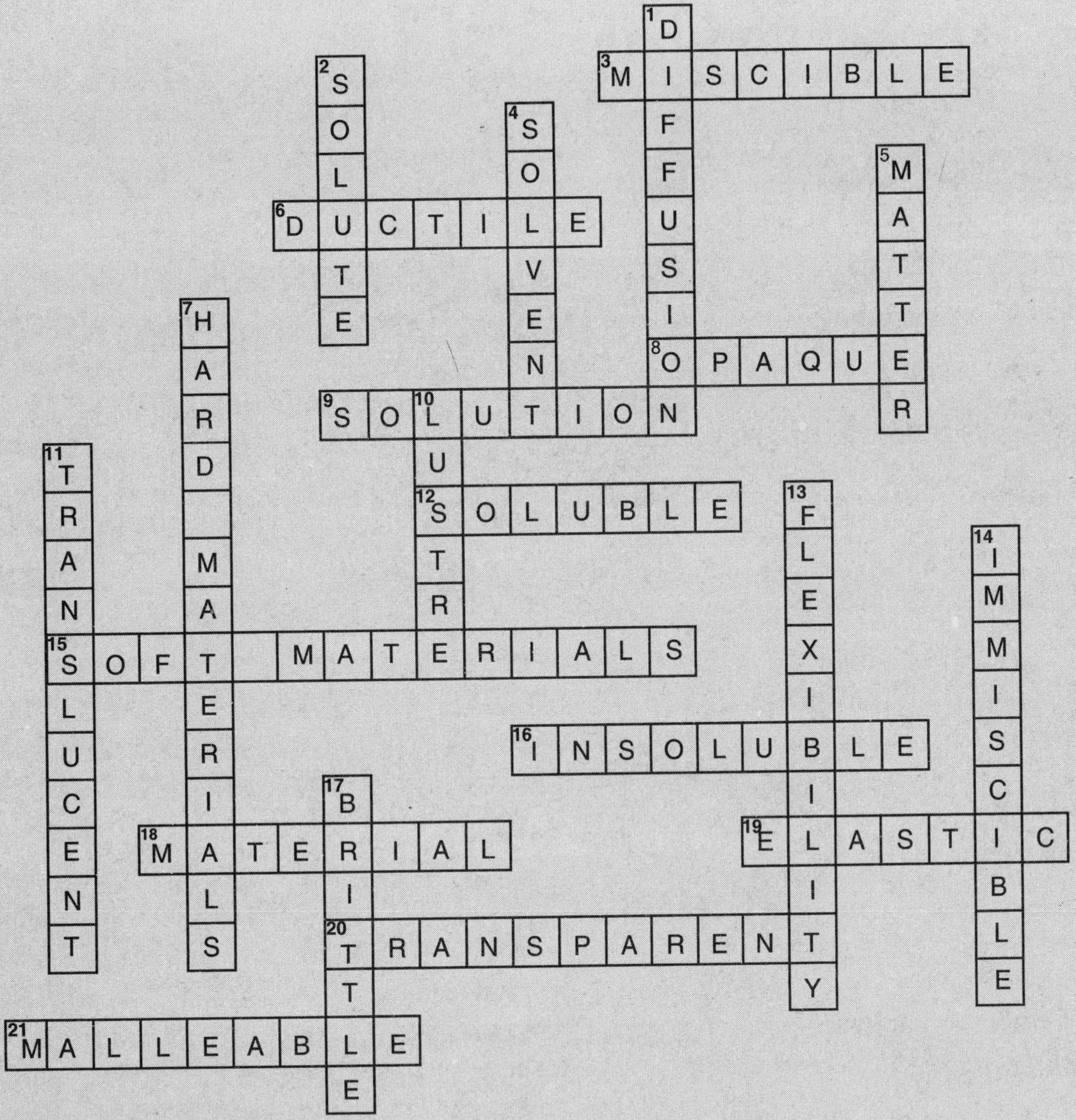

Chapter

# 8 SEPARATION OF SUBSTANCES

## KEY FACTS

**1. Mixture**. A mixture is a combination of two or more different substances. The substances which make up the mixture are called **components** or **constituents** of the mixture. These components retain their original properties and can be separated easily from each other.

Some of the mixtures around us are air, sea-water, tea, petroleum etc.

(i) Air is a mixture of gases.

(ii) Sea-water is a mixture of water, common salt, and many other salts dissolved in it.

(iii) A cup of tea is a mixture of water, tea leaves extract, sugar and milk.

(iv) Petroleum (crude oil) is a mixture of various carbon compounds.

**2. Need to separate different components of a mixture**. We may have to separate different components because of several reasons. Some of the reasons may be

(i) to remove undesirable components like removal of tea-leaves when preparing tea.

(ii) to separate the harmful components like removing stones and insects from rice and pulses etc., before cooking them.

(iii) to obtain the pure sample of a substance, like removing dissolved salts by distillation from tap water to obtain pure water for medicinal purpose.

(iv) to obtain useful components–like separating useful substance like kerosene, petrol and diesel from petroleum mixture.

**3. Mixtures may be homogenous, or heterogenous**

(i) Mixtures like syrup, sugar water in which the components are thoroughly mixed and are not seen as separate substances are called **homogenous mixtures**.

(ii) Mixtures like that of wheat and husk, sand and water, in which different components can be easily seen are called **heterogenous mixtures**.

**4. Methods of separating components of mixtures.**

(1) **Threshing.** This method is used to separate grains from the stalks. The stacks of harvested crop are held in hand and hit on a hard surface.

(2) **Winnowing.** This method is used to separate the heavier components from the lighter ones, like heavier wheat grains from the lighter husk. The farmer takes the mixture of husk and grains in a winnow basket and stands on a high platform. He now releases the mixture by shaking the winnow basket continuously. The wheat grains, being heavy, fall down vertically to the ground while the husk particles, being lighter are blown off by wind a little away.

(3) **Hand-picking.** This method is generally used to remove undesirable substances like stones, insects, etc., from wheat, rice and pulses.

(4) **Sieving.** In this method, the fine particles of the mixture pass through the holes of the sieve while the bigger particles remain on the sieve. Thus, a sieve may be used to remove bran from the flour and fine grains of sand from pebbles required to mix with cement.

(5) **Magnetic separation.** Magnets are used to clean up agricultural products. For example, sugar on conveyor belts are passed near magnets placed above it so that any pieces of iron such as metal chips from machinary in the sugar can be removed.

Magnetic separation may be used in the treatment of sewage. When magnetic grains are mixed with sewage, the tiny pieces of sewage stick to the grains. When the sewage passes through strong magnets, the mixture of magnetic grains and sewage are attracted to the magnets, leaving almost clean water to be released into the ocean. About 90% of the solid sewage is removed in this way and at a much faster rate. The magnetic grains are then separated from the waste sludge and reused.

(6) **Sedimentation and decantation.** This method is used to remove insoluble solid particles from a liquid. Suppose, we have a mixture of soil and water. It is taken in a utensil and allowed to stand for sometime. After some time, the sand and mud would settle down at the bottom of the utensil. The settled sand and mud at the bottom of the utensil are called **sediments** and this process of settling down of heavier and insoluble components of the mixture is called **sedimentation.**

The clean liquid (water) over the sediments is then poured into another utensil, without disturbing the sediments. The process of transferring elean liquid is called **decantation.**

Decantation process is used by housewives while washing vegetables, rice, and pulses. The food item settles down in the washing utensil and the dust and other light weight impurities are taken out with the decanted water.

(7) **Loading.** This method is used to make sedimentation faster. A crystalline piece of alum is put on the surface of muddy water. Some alum gets dissolved in water. Its particles get attached to the dust particles making them heavy in weight. This results in their rapid settling down.

(8) **Filtration.** In the process of filtration, insoluble solid particles are trapped in the filter as **residue** while the liquid passes through the filter and is collected as **filtrate.**

(9) **Evaporation.** The changing of a liquid into vapours (or gas) is called evaporation. It is used to separate dissolved solid substance from a liquid like sugar or salt from water. When a solution is heated, the liquid or solvent in the solution evaporates leaving behind the dissolved solids as residue.

(10) **Distillation.** Distillation is used to separate a solvent from a solution. In this process, the solution is heated so that its liquid component evaporates and escapes as a vapour. The vapour is then cooled and condensed into a liquid called the **distillate.**

Distillation is widely used in the perfume industry. It is also used to obtain pure water from seawater in some countries such as Saudi Arabia.

(11) **Churning.** It is a common method being used in our homes to obtain butter from curd. In a dairy, cream is separated from milk with the help of churning machine.

(12) **Sublimation.** The process of changing from a solid to a gas, or from a gas to a solid, without turning into a liquid first is called **sublimation**. Frost sometimes evaporates by sublimation.

The property of sublimation is also sometimes used to separate certain substances from a mixture in which they may be present. For example, if ammonium chloride gets accidentally mixed up with common salt, then to separate it, the mixture is warmed in a flask. Ammonium chloride changes to vapour which cools to solidify at the neck of the flask. Salt is left behind.

(13) **Multiple separation techniques** may have to be applied sometimes if a mixture has more than two components. For example, if we have a mixture of salt and sand, then water is first added. Salt gets dissolved in water and sand is separated by the process of filtration. Now the salt solution may be put in the sun or heated. Water will evaporate, leaving behind the crystals of salt.

(14) **Saturated solutions.** A given quantity of water can only dissolve upto a certain amount of a substance at a particular temperature. The solution which contains the maximum possible amount of the substance which can be dissolved in it is called a **saturated solution**.
The measure of how much of a substance can dissolve in another substance is called **solubility**. It depends on the two substances involved and the temperature. For example,
In particular, **the maximum amount of a substance which can be dissolved in 100 grams of water at a given temperature is known as the solubility of that substance in water at that temperature.**
Thus, the solubility of sugar in water is 40 grams at 20°C because this is the maximum amount of sugar that can be dissolved in 100 g of water at 20°C.

## Question Bank–8

**1.** What methods will you use to obtain clear water from a sample of muddy water?

**2. Fill in the bianks with suitable words.**
(a) The method of separating seeds of paddy from its stalks is called ________.
(b) When milk, cooled after boiling is poured onto a piece of cloth, the cream (*malai*) is left behind on it. This process of separating cream from milk is an example of________.
(c) Salt is obtained from seawater by the process of ________.
(d) Impurities settled at the bottom when muddy water was kept overnight in a bucket. The clear water was then poured off from the top. The process of separation used in this example is called ________.
(e) The process of heating a liquid to form vapour, and then cooling the vapour to get back liquid is called ________.
(f) Kerosene from oil can be separated by making use of a ________.

**3.** What is condensation?
(a) change of gas into solid
(b) change of solid into liquid
(c) change of vapour into liquid
(d) change of heat energy into cooling energy

**4. Match the following.**

| **Process** | **Changes** |
|---|---|
| A. Evaporation | 1. Liquid into gas |
| B. Sublimation | 2. Gas into liquid |
| C. Freezing | 3. Solid into gas |
| D. Melting | 4. Solid into liquid |
| | 5. Liquid into solid |

A B C D　　　　A B C D
(a) 1 2 5 4　　(b) 3 1 2 4
(c) 3 1 5 4　　(d) 1 3 5 4

**5.** Which of the following exhibit the property of sublimation?
(a) Ice　　(b) Wax
(c) Camphor　　(d) Common Salt

**6.** A mixture of ammonium chloride and sand is separated by
(a) Evaporation　　(b) Decantation
(c) Filtration　　(d) Sublimation

**7.** A solid substance is dissolved in water. Which one of the following methods is used to separate it?
(a) Filtration　　(b) Evaporation
(c) Sublimation　　(d) Decantation

**8. Match the following :**

| | |
|---|---|
| (i) Separation of butter and curd | (a) Loading |
| (ii) Separation of chaff from wheat flour | (b) Magnet |
| (iii) Only salt from salt solution | (c) Decantation |
| (iv) Salt as well as water from salt solution | (d) Churning |
| (v) Making clay particles of muddy water heavier by depositing alum on them | (e) Distillation |

(vi) Separation of chalk from water solution (f) Filtration
(vii) Tea leaves from tea (g) Evaporation
(viii) Separation of iron filings and sulphur (h) Sieving
(ix) Sand water

**9. State whether the following statements are true or false.**

(i) The process of winnowing is used to remove small stone particles from wheat.
(ii) A mixture of powdered salt and sugar can be separated by the process of winnowing.
(iii) Grain and husk can be separated by the process of decantation.
(iv) Cream (malai) is separated from milk by sieving.
(v) A mixture of milk and water can be separated by filtration.
(vi) Separation of sugar from tea can be done by filtration.
(vii) Common salt is separated from its solution in water by decantation.
(viii) Solubility of a substances decreases on lowering the temperature.

**10.** Lemonade is prepared by mixing sugar and lemon juice in water. You wish to add ice to cool it. Should you add ice to the lemonade before or after dissolving sugar in water? In which case it would be possible to dissolve more sugar?

**11.** A mixture of tea leaves and iron filings can be separated by:
(a) filtration (b) hand picking
(c) magnet (d) sieving

**12.** You can separate salt from water by evaporation. Can you separate sugar also from a sugar solution by this method?

**13.** You can remove iron filings from sand by magnetic separation method. This method can be applied only when
(a) magnetic substances are mixed up with non-magnetic substances
(b) magnetic substances are mixed up with magnetic substances

**14.** The process of settling down of heavier and insoluble components in a mixture solution is called
(a) filtration (b) sieving
(c) sedimentation (d) sublimation

**15.** A dishonest shopkeeper mixes small pebbles into a bag of rice. What method will you use to separate the pebbles from the mixture?
(a) Hand picking (b) Sieving
(c) Winnowing
(d) Magnetic separation

**16.** Cooking oil can be converted into vegetable ghee by the process of
(a) oxidation (b) hydrogenation
(c) distillation (d) crystallisation

**17.** A mixture of iron filings and sand can be separated by
(a) Heating (b) Sublimation
(c) Hand picking
(d) Magnetic separation

**18.** Sieving can be used only when the components of the mixture have
(a) different sizes
(b) different melting points
(c) different boiling points
(d) different densities

**19.** A mixture of two or more gases can be separated by
(a) Sedimentation (b) Liquefication
(c) Distillation (d) Decantation

**20.** A bag contains a mixture of two solids *A* and *B*, solid *A* being lighter than solid *B*. Solid *A* is an impurity which needs to be removed. What type of separating method will you use to obtain pure solid *B*?
(a) Sieving
(b) Threshing
(c) Winnowing
(d) Magnetic separation

**21.** Decantation can be used to separate
(a) sugar from sugar solution (solid dissolves in the liquid)
(b) two miscible liquids, *i.e.*, which mix with

each other, like alcohol and water.

(c) two immiscible liquids, *i.e.*, liquids which do not mix together like oil and water

(d) solid–solid mixture

**22.** Alum particles are deposited on suspended clay particles of muddy water to make them heavy so that they settle down rapidly. The process is called

(a) Sedimentation (b) Filtration

(c) Loading (d) Sieving

**23.** What common property do iodine, camphor, ammonium chloride and napthalene have?

(a) Equal melting points (b) Equal densities

(c) Sublimation (d) None of these

**24.** Conversion of a substance directly from solid to vapour state is known as

(a) distillation (b) evaporation

(c) vapourisation (d) sublimation

**25. Find the odd one out :**

(a) filtration, evaporation, loading, crystallisation

(b) solutions, compounds, suspensions, emulsions

(c) soil, air, sea-water, table-salt, milk

(d) solution, filtrate, emulsion, residue

**26.** How will you separate the following?

(a) A mixture of sand and salt

(b) Sugar mixed with wheat flour

(c) Mixture of iron filings, chalk powder and common salt

(d) Common salt, sulphur, iodine and iron filings.

**27. Complete the following concept map :**

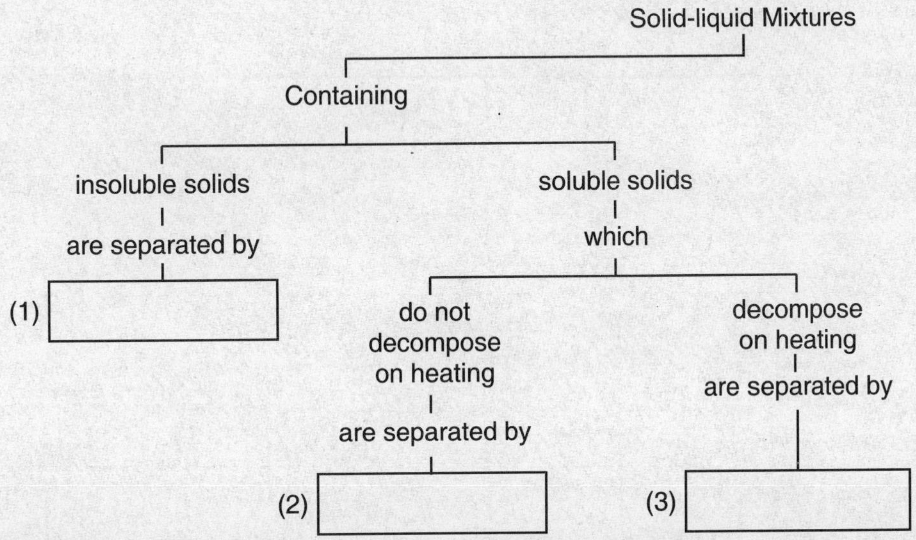


## Answers

**1.** Sedimentation, Decantation, Filtration

**2.** (a) threshing (b) filtration (c) evaporation (d) decantation (e) distillation (f) separating funnel

**3.** (c) **4.** (d) **5.** (c)

**6.** (d) **7.** (b)

**8.** (i) – (d), (ii) – (h), (iii) – (g), (iv) – (e), (v) – (a), (vi)– (b), (vii) – (f),(viii) – (b), (ix) – (c)

**9.** (i)False (ii) False (iii) False (iv)False (v) False (vi) False (vii) False (viii) True

**10.** Ice should be added after dissolving sugar in water because the solubility of a substance decreases with decrease in temperature and so it will be possible to dissolve more sugar before adding ice.

**11.** (c)

**12.** No, because sugar decomposes upon heating. Crystallisation is used to separate a soluble solid that decomposes on heating from its solutions.

| | | |
|---|---|---|
| **13.** (a) | **14.** (c) | **15.** (a) |
| **16.** (b) | **17.** (d) | **18.** (a) |
| **19.** (b) | **20.** (c) | **21.** (c) |
| **22.** (c) | **23.** (c) | **24.** (d) |

**25.** (a) loading (b) compounds (c) table-salt (d) residue

**26.** (a) Add water. Salt gets dissolved in water. Separate it by filtration. Remove water from the filtrate (salt solution) by evaporation, leaving behind crystals of salt.

(b) Add water and stir. Sugar dissolves in water forming sugar solution whereas flour remains undissolved. Remove it by filtration. Now obtain pure sugar by evaporation.

(c) (i) Remove iron filings by using a magnet.

(ii) Add water. Salt dissolves while chalk powder does not dissolve. Remove chalk powder by filtration.

(iii) Now obtain common salt by evaporation.

(d) (i) Use magnet to remove iron filings.

(ii) Add water. Salt is dissolved in water. Obtain salt by filtration and crystallisation.

(iii) Iodine is separated from the mixture by the process of sublimation. Sulphur is left behind.

**27.** (1) Filtration

(2) Evaporation

(3) Crystallisation

*Chapter*

# 9 CHANGES AROUND US

## KEY FACTS

1. Changes are taking place all the time around us.
   **For example**, a change in weather from hot to cold and vice-versa, flowering of plants, change of water into steam or ice, formation of curd from milk, burning of fuels etc.
2. A change may involve different kinds of alterations in a thing like, change in position, shape, size, colour, temperature, composition and structure. A kick changes the position of a football. Heating changes water to vapour or steam.
3. A change cannot occur on its own. There is always a cause which brings about a change. Thus, heat is the cause which turns ice into water.
4. Some changes are beneficial to us whereas some are harmful. Thus, the ripening of fruits is a beneficial change whereas the rottening is a harmful change.
5. **A. All changes in matter involve energy**. Heat energy causes ice to melt. The energy of a moving ball breaks glass. This energy comes from the muscles of the person who hit or threw it. Energy is either given off or taken in.

**Classification of changes**

6. **Physical and chemical changes.**

**A.** A **physical change** is a change in the way matter looks. Without changing it into a new kind of matter.
**For example**.
(a) When a ball hits a window pane, the glass may shatter into hundreds of tiny pieces. Breaking the window causes a change in the shape and size of the window glass, but each tiny piece of glass still has the properties of glass.
(b) When a balloon is blown up, it looks bigger, but it is still made of the same kind of matter. A change in size is a physical change.

**B. A chemical change** is a change in matter that produces new kinds of matter with different properties. Energy is always involved in a chemical change. It is either given off or taken in.
**For example,** burning is a chemical change that gives off light and heat, which are two forms of energy. In a physical change, particles (atoms and molecules) rearrange but in a chemical change, particles break up and form new combinations with other atoms and molecules.
A **chemical reaction** is another term for a chemical change. In a chemical reaction, the matter that you start with is called the **reactant** and the newly formed matter is called the **product**.
For example, when you make cake, the flour, sugar, butter, chocolate, etc., are the reactants and the cake is the product.

**C. Slow and fast changes** - Burning of a piece of paper when brought over the flame of a candle and a matchstick catching fire immediately on striking against a rough surface are examples of fast changes whereas curdling of milk is a slow change.

**D. Reversible and Irreversible changes**

(1) The changes that can be reversed are called **reversible changes**. A physical change is a reversible change.

**For example,**

(a) the stretching of rubber band is a reversible change. The stretched rubber band returns to its original state when released.

(b) dissolving salt in water is a reversible change because water can be evaporated by heating and salt is left behind.

(c) the melting of coal tar on heating is a reversible change because on cooling the hot molten tar solidifies again.

(2) The change that cannot be reversed to form the original substance or substances is called **irreversible change**. A chemical change is an irreversible change. Irreversible changes are permanent.

**For example,**

(a) Wheat once converted into flour cannot be reconverted into wheat. It is an irreversible change.

(b) The growth of a child to become a youngman is an irreversible change because the youngman cannot be changed to become a child again.

(c) The burning of paper is an irreversible change because the ash and smoke produced on burning the paper cannot be reconverted into paper.

**E. Periodic and non-periodic changes.**

(1) The changes which take place at fixed time intervals are called **periodic changes**.

**For example,** changing of day into night and of night into day take place at fixed intervals of time and so is a periodic change.

(2) The changes which may take place at any time and not at fixed intervals are called **non-periodic changes**. Thus, change in weather, break-down in electric supply, happening of an accident are all non-periodic changes.

**F. Desirable and undesirable changes.**

(1) Those changes which are useful and we want them to happen are **desirable** changes.

**For example,** ripening of fruits, blooming of flowers, digestion of food, water cycle in nature are all desirable changes.

(2) Those changes which are harmful to us and which we don't want to happen are called **undesirable changes**.

Thus, spoilage of food, landslides, an earthquake, a cyclone, a cloud burst are all undesirable changes.

**G. Endothermic and Exothermic changes.**

(1) **An endothermic change** is the change in which heat from outside is absorbed by the reactants, leaving the surroundings cooler. Examples are evaporation of water, alcohol, nail polish remover and dissolving glucose in water.

(2) **An exothermic change** is the change in which heat is given out during the course of a reaction. Examples are burning of wood, coal, petrol and other fuels, the process of breathing and dissolving of washing soda or quick lime in water.

## Question Bank–9

**1.** How do you know that melting of ice is a physical change?

**2.** A plumber uses a torch to heat a copper pipe in order to bend it. Is this change a physical change? Explain.

**3.** Compare the properties of broken and unbroken glass. Explain why breaking glass is a physical change.

**4.** Which of the following is NOT a physical change?
(a) A dish of water kept out in the sun becomes dry.
(b) Wood burns in a campfire to form charcoal.
(c) A sandwich is cut into four smaller pieces.
(d) Chocolate melts in your pocket on a hot day.

**5.** How are new kinds of matter formed in a chemical change?

**6.** When you add vinegar to baking soda, you watch it bubble over and feel the container get slightly cooler. Is this a chemical or physical change? Explain.

**7.** Which of the following is a chemical change?
(a) bending a paper clip
(b) mixing water and salt
(c) rusting iron
(d) freezing water

**8.** The growth of seedling into a plant is
(a) undesirable change
(b) desirable change
(c) reversible change
(d) physical change

**9.** Which of the following statements is true?
(a) Chemical changes are reversible
(b) Growth of a tree is a periodic change
(c) Rusting of iron is a fast change
(d) Energy is required to cause a change

**10.** Classify the following changes into fast and slow changes.
(1) germination of a seed
(2) explosion of a cracker
(3) growth of a plant
(4) fading of colours
(5) taking photograph by exposing a film in a camera
(6) tooth decay

**11.** To walk through a waterlogged area, you usually shorten the length of your dress by folding it. Can this change be reversed?

**12.** You accidentally dropped your favourite toy and broke it. This is a change you did not want. Can this change be reversed?

**13.** The germination of a seed is
(a) a physical, reversible, desirable change in which energy is produced
(b) a fast, chemical reaction in which energy is released
(c) a slow irreversible, desirable, chemical change
(d) a fast, irreversible, undesirable chemical change

**14.** Some changes are listed below. For each change write whether the change can be reversed or not. **(Yes/No).**
1. Sawing of a piece of wood
2. Melting of ice candy
3. Dissolving sugar in water
4. Cooking of food
5. Ripening of a mango
6. Souring of milk
7. Drawing a picture on a drawing sheet.
8. Lighting a matchstick
9. Heating a piece of iron-wire to red hot
10. Decay of food, dead plant and animal bodies.

**15.** In a change, energy is
(a) absorbed
(b) released
(c) either absorbed or released
(d) None of these

**16.** Changes involve
(a) an action (b) a reaction
(c) interactions (d) none of these

**17.** What is common among the following changes.
(1) Planet's revolution around the sun
(2) Blinking of traffic lights, occurence of solar eclipse.
(a) Irreversible changes
(b) Chemical changes
(c) Periodic changes
(d) Undesirable changes.

**18.** How do the following changes differ from one another?
(a) Melting of wax (b) Burning of wax

**19.** Which of the following is a physical change?
(a) Boiling of water
(b) Boiling of an egg
(c) Burning of wood
(d) Blooming of a flower

**20.** Which of the following is a periodic change?
(a) Stoppage of electric supply
(b) Occurance of low and high tides
(c) Falling of ripened fruits from a tree
(d) Sneezing

**21.** The gas we use in the kitchen is called liquified petroleum gas (LPG). In the cylinder it exists as a liquid. When it comes out from the cylinder it becomes a gas (change-*A*) then it burns (change-*B*). The following statements pertain to these changes. **Choose the correct one.**
(a) Process *A* is a chemical change.
(b) Process *B* is a chemical change.
(c) Both process *A* and *B* are chemical changes.
(d) None of these processes is a chemical change.

**22.** Anaerobic bacteria digest animal waste and produce biogas (change-*A*). The biogas is then burnt as fuel (change-*B*). The following statements pertain to these changes. Choose the correct one.
(a) Process *A* is a chemical change
(b) Process *B* is a chemical change
(c) Both processes *A* and *B* are chemical changes
(d) None of these processes is a chemical change.

**23.** Which of the following statements pertaining to a chemical change are true?
(a) New substances are formed
(b) Irreversible
(c) Either exothermic or endothermic
(d) All of the above

**24.** If few drops of petroleum are poured on your palm, you feel cool as the petrol evaporates. This change is
(a) slow change
(b) chemical reaction
(c) endothermic change
(d) periodic change

**25.** Which one of the following is a chemical change?
(a) Melting of ice
(b) Burning of coal
(c) Magnetizing of iron
(d) Reducing a solid stone to a fine powder

**26. State whether the following statements are true or false:**
(1) The softening of iron on heating to red hot stage is a reversible change.
(2) Ageing of humans and animals is a reversible change.
(3) Appearance of Halley's Comet once in every 76 years is a periodic change.
(4) Forest fire is a desirable physical change.
(5) When petrol is burnt, flames are produced. This is an endothermic reaction.
(6) When baking soda is mixed with lemon juice bubbles are formed with the evolution of a gas. It is a physical change.
(7) When a candle burns both the physical and chemical changes take place.
(8) Condensation of steam is not a chemical change.
(9) Cutting of a log of wood is a reversible physical change.
(10) Liquefaction of air is a chemical change.
(11) Iron and rust are the same substances.
(12) Breaking down of ozone (a gas) is a chemical change.
(13) Corrosion of metal is a physical change.
(14) Explosion is a chemical change.
(15) Occuring of the full moon and new moon is a periodic change.
(16) Formation of clouds is a slow, desirable and reversible physical change.

**27.** Photosynthesis is
(a) an exothermic process
(b) an endothermic process
(c) a neutral process
(d) a thermostatic process

**28.** Which of the following is a chemical change?
(a) Magnetisation of iron (b) Melting of ice
(c) Burning of sulphur (d) Melting of wax

**29.** Which of the following substances undergoes chemical change an heating?

(a) Sodium chloride (b) Silica

(c) Lead nitrate (d) Platinum wire

**30.** Chemical change does not take place in case of

(a) formation of manure from leaves

(b) beating aluminium to make aluminium foil

(c) digestion of food

(d) burning of magnesium ribbon in air

**31.** If a weight is suspended on a spring, it would stretch. This change is

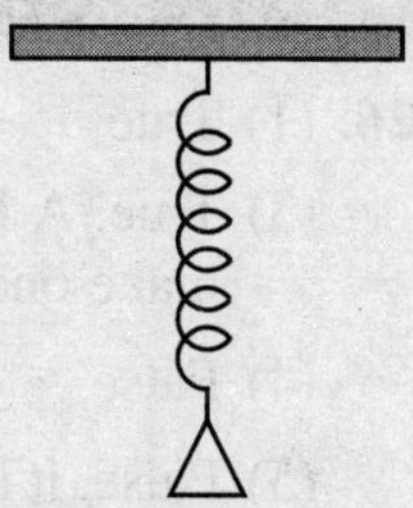

(a) man-made

(b) reversible

(c) irreversible

(d) both (a) and (b)

**32.** Study the Venn diagram and tell which of the following changes are represented by *A*, *B*, *C* and *D*.

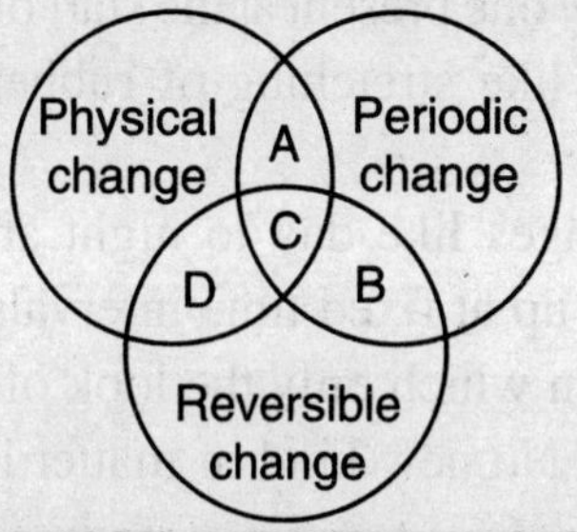


1. Occurence of seasons
2. Dissolving sugar in water
3. Solar eclipse
4. Motion of blades of a fan rotating at uniform speed.

**33.** Which of the following is a physical change?

(a) Oxidation (b) Reduction

(c) Sublimation (d) Decomposition.

**34. Solve the following crossword with the help of the given clues :**

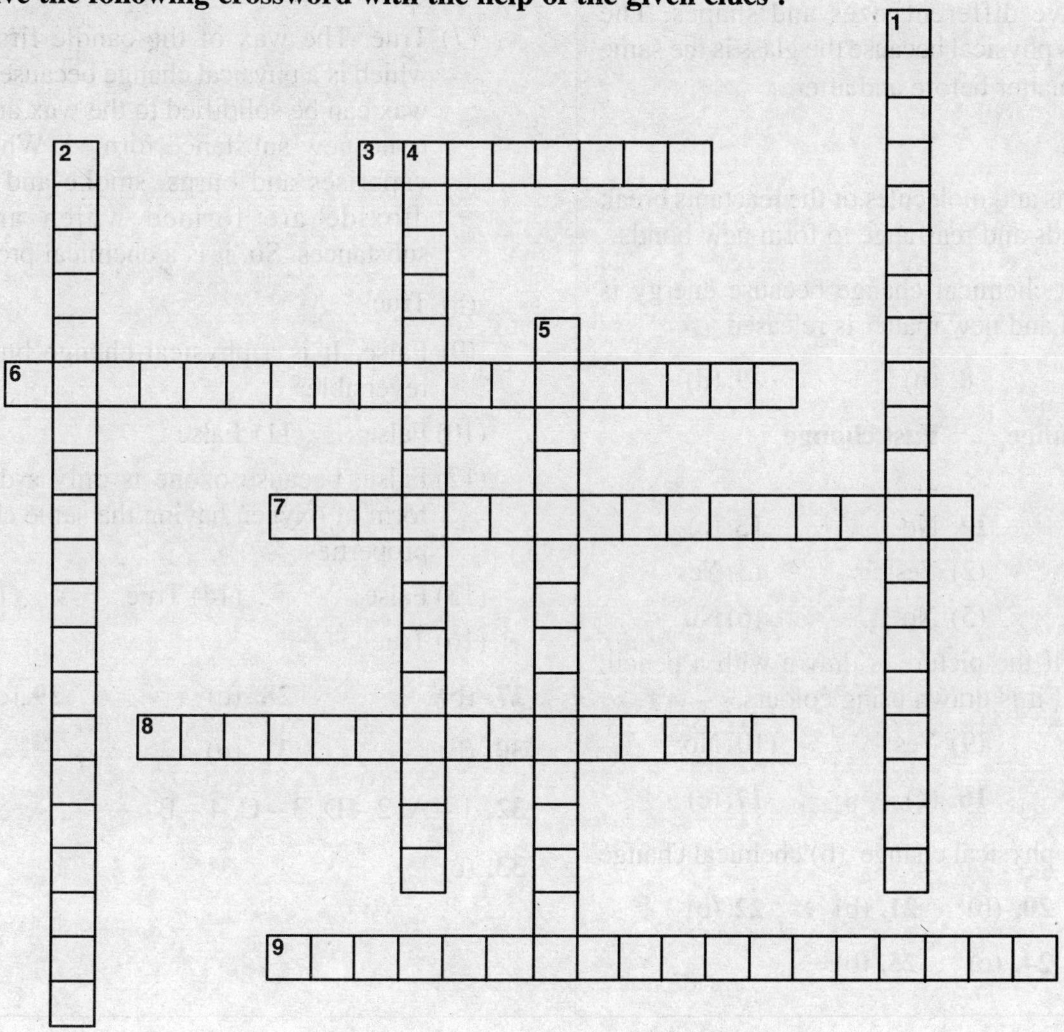

**ACROSS**

3. A substance participating in a chemical reaction, especially one present at the start of the reaction.
6. Changes like stretching of rubber that can be reversed.
7. The changes like day to night and vice-versa that take up at fixed time intervals.
8. Change in which only the look of the matter is changed. No new kind of matter is produced.

**DOWN**

1. Changes like conversion of wheat into flour which cannot be reversed.
2. Changes which can take place at any time and not at fixed intervals.
4. Change in which heat in released.
5. Change in which new kind of matter is produced.
9. Change in which heat is absorbed.

## Answers

**1.** Melting ice is a physical change because the ice and the liquid water are both forms of water. No new matter is formed during the change.

**2.** The bending of copper is a physical change because the matter is still copper after the change.

**3.** Broken and unbroken glass share the properties of transparency, smoothness and sharp edges. They have different sizes and shapes. The change is physical because the glass is the same kind of matter before and after.

**4.** (b)

**5.** The atoms and molecules of the reactants break their bonds and rearrange to form new bonds.

**6.** This is a chemical change because energy is absorbed and new matter is released.

**7.** (c) **8.** (b) **9.** (d)

**10.**

| **Slow change** | **Fast change** |
|---|---|
| 1, 3, 4, 6 | 2, 5 |

**11.** Yes **12.** No **13.** (c)

**14.** (1) No (2) Yes (3) Yes
(4) No (5) No (6) No
(7) Yes, if the picture is drawn with a pencil. No, if it is drawn using colours.
(8) No (9) Yes (10) No

**15.** (c) **16.** (c) **17.** (c)

**18.** (a) It is a physical change (b) chemical change

**19.** (a) **20.** (b) **21.** (b) **22.** (c)

**23.** (d) **24.** (c) **25.** (b)

**26.** (1) True (2) False
(3) True. A Halley's Comet takes 76 years to make one complete orbit around the sun.
(4) False
(5) False. It is an exothermic reaction in which heat energy is released.
(6) False. A new substance carbon dioxide is formed and so it is a chemical change.
(7) True. The wax of the candle first melts which is a physical change because molten wax can be solidified to the wax and there is no new substance formed. When wax vaporises and burns, smoke and carbon dioxide are formed which are new substances. So, it is a chemical process.
(8) True
(9) False. It is a physical change but is not reversible
(10) False (11) False
(12) False, because ozone is only a different form of oxygen having the same chemical properties.
(13) False (14) True (15) True
(16) True

**27.** (b) **28.** (c) **29.** (c)

**30.** (b) **31.** (d)

**32.** 1 – A, 2 – D, 3 – C, 4 – B

**33.** (c)

**34.**

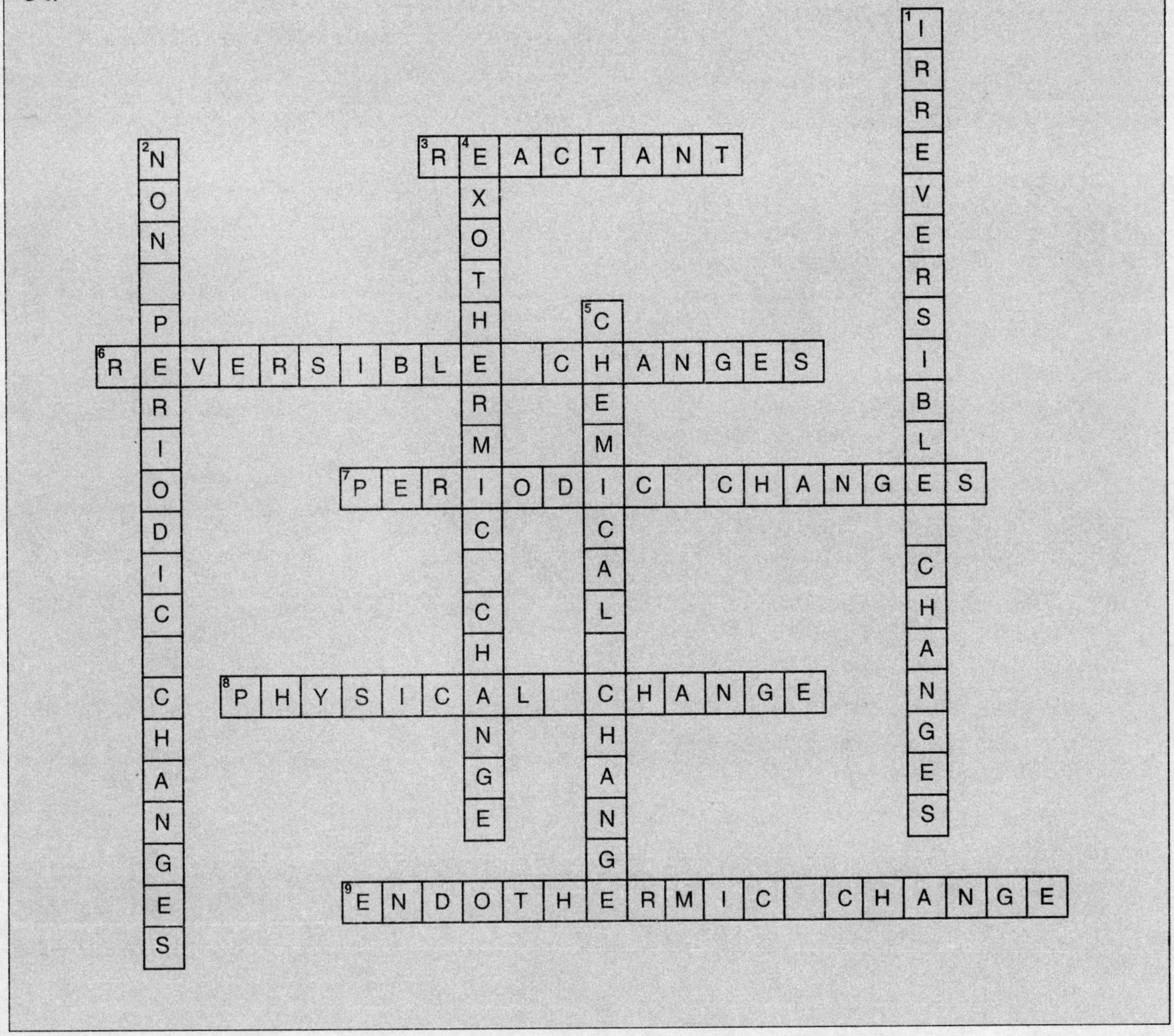
1 IRREVERSIBLE CHANGES
2 NON PERIODIC CHANGES
3 REACTANT
4 EXOTHERMIC CHANGE
5 CHEMICAL CHANGE
6 REVERSIBLE CHANGES
7 PERIODIC CHANGES
8 PHYSICAL CHANGE
9 ENDOTHERMIC CHANGE

# Chapter 10

# WATER AND AIR

## KEY FACTS

### WATER

1. Water exists in three states – Solid, liquid and gas. These states are interchangeable.
   **Ex.** Solid $\rightarrow$ ice, liquid $\rightarrow$ water, gas $\rightarrow$ vapour or steam.
2. Water changing into water vapour on heating is called **evaporation**.
3. The process of conversion of vapour into liquid form on cooling is called **condensation**.
4. Condition of extreme dryness due to lack of rain for a long period is called **drought.**
5. An arrangement to collect and store rain water for future use is called **rainwater harvesting**.
6. Plants absorb water from the soil with the help of their roots and use a part of it in making food. A major part of this water is lost to the air. The loss of water from plants as water vapour through the pores of their leaves is called **transpiration**.
7. **Water cycle** is the journey of the water from the oceans and the land to the atmosphere and back to the earth.

### AIR

1. The **atmosphere** is the envelope of air surrounding the earth.
2. Air in the atmosphere is a mixture of gases. It includes **nitrogen, oxygen, carbon dioxide, water vapour, dust particles** and some gases in small ratio.
3. Nearly 78% of the air is nitrogen, 21% oxygen and the rest 1% includes carbon dioxide, water vapour, dust particles and some other gases.
4. The composition of air can change from place to place.
5. Oxygen is essential for life and combustion.

## Questions Bank–10

1. What does hail consist of ?
   (a) Granular ice
   (b) Crystals of ice
   (c) Water droplets
   (d) Masses of ice in layers one above the other.
2. Release of water through leaves of plants is known as
   (a) Photosynthesis (b) Respiration
   (c) Evaporation (d) Transpiration
3. **State for each of the following whether it is due to evaporation or condensation :**
   (a) Water drops appear on the outer surface of a glass containing cold water.
   (b) Steam rising from wet clothes while they are ironed.
   (c) Fog appearing on a cold winter morning.
   (d) Blackboard dries up after wiping it.
   (e) Steam rising from a hot girdle when water is sprinkled on it.

**4. Which of the following statements are "true".**

(a) Water vapour is present in air only during the monsoon.

(b) Water evaporates into air from oceans, river and lakes but not from the soil.

(c) The process of water changing into its vapour, is called evaporation.

(d) The evaporation of water takes place only in sunlight.

(e) Water vapour condenses to forms tiny droplets of water in the upper layers of air where it is cooler.

**5.** The process of removing salt from sea water is called

(a) Desalination (b) Dehydration

(c) Decantation (d) Re-hydration

**6. Fill in the blanks in the flow chart to complete the water cycle.**

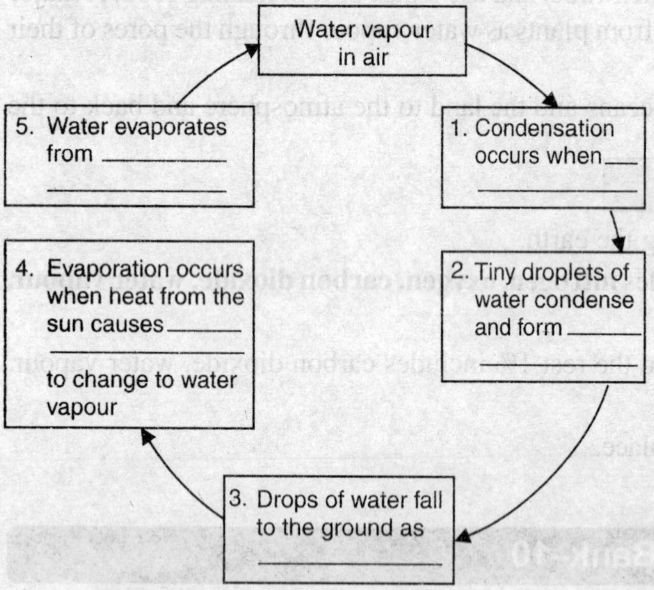


**7.** A storm is predicted if atmosphere pressure

(a) Rises suddenly (b) Rises gradually

(c) Falls suddenly (d) Falls gradually

**8.** The gas which turns into liquid at the lowest temperature among the following is —

(a) Hydrogen (b) Oxygen

(c) Helium (d) Nitrogen

**9.** Which of the following celestial bodies contains abundant quantities of Helium - 3 a potential source of energy?

(a) Earth (b) Moon

(c) Venus (d) Saturn

**10.** A soft plastic ball filled with air at atmospheric pressure weighs $W_1$. If weight becomes $W_2$ when air is removed from it. Then,

(a) $W_1 > W_2$ (b) $W_1 = W_2$

(c) $W_1 < W_2$ (d) $W_1 = 2W_2$

**11.** Recharging of water table depends on

(a) amount of rainfall

(b) relief of the area

(c) vegetation of the area

(d) amount of percolation

**12.** The temperature of the air varies

(a) Periodically

(b) Vertically

(c) Horizontally

(d) All are correct

**13.** The relative humidity of a place is measured by the

(a) Amount of sunshine

(b) Amount of water vapour in the air

(c) Amount of the rainfall in the past 24 hours

(d) Strength of the wind

**14.** The Ozone layer of the earth is useful for living beings because

(a) it serves as the source of oxygen by decomposing air

(b) it maintains the nitrogen cycle of the earth

(c) it maintains the temperature of the earth

(d) it protects them from excessive ultraviolet rays of the sun

**15.** Which one of the following gases present in the upper atmosphere cuts off a major portion of ultraviolet radiation of the sun from reaching the earth?

(a) Carbon dioxide (b) Ozone

(c) Nitrogen (d) Oxygen

**16.** Use the diagram below to answer questions (i) to (iii).

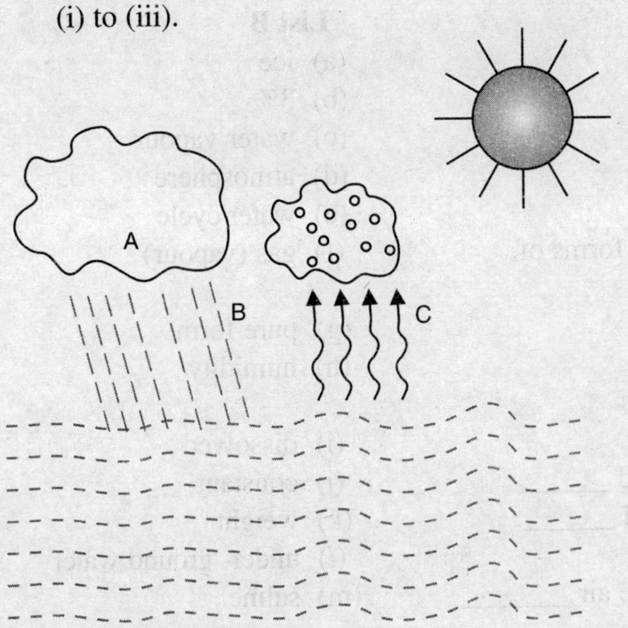


(i) What happens at A?

(ii) What happens at B?

(iii) What happens at C?

**17.** A trail is created by jet-planes as they release —through the exhaust.

(a) Carbon-dioxide (b) Water vapours
(c) Carbon monoxide (d) Hydrogen

**18.** Water boils at a lower temperature than 100°C on a hill station because

(a) Water vapours are less at high altitudes
(b) Temperature is lower at high altitudes
(c) Pressure is lower at high altitudes
(d) There is cloud formation at high altitudes

**19.** The most prominent gases in the atmosphere in terms of volume are

(a) Nitrogen and methane
(b) Oxygen and nitrogen
(c) Hydrogen and nitrogen
(d) Oxygen and carbon dioxide

**20.** Which one of the following is the most abundant gas in the earths' atmosphere other than nitrogen and oxygen?

(a) Argon (b) Carbon dioxide
(c) Hydrogen (d) Methane

**21.** The air consists of 79.2% of Nitrogen, 20.7% of Oxygen, 0.08% of other light gases and the remaining gas is Argon. Find out the volume of the air consisting of one cubic metre of Argon.

(a) 500 $m^3$ (b) 50 $m^3$
(c) 5 $m^3$ (d) 5000 $m^3$

[**Hint.** Volume of Argon in air

$$= 100 - (79.2 + 20.7 + 0.08) = 0.02\%$$

$\therefore$ Volume of air for 1 $m^3$ of Argon

$$= \frac{100}{0.02} = \frac{100\times100}{2} = 5000\ m^3]$$

**22.** Use the graph below to answer the following questions.

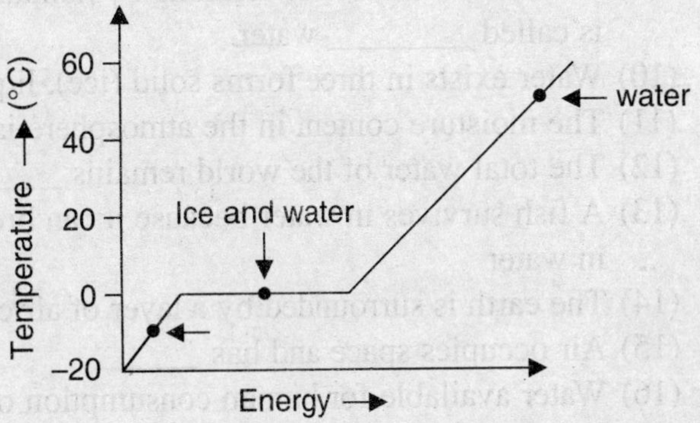


(i) The graph shows that the temperature remains unchanged until all the ice has ______.

(ii) If energy were removed from the water, the temperature would remain unchanged until all the liquid had ______.

**23.** Life first originated in

(a) Air (b) Space
(c) Water (d) Soil

**24.** Water is a liquid between

(a) 0°C to 90°C (b) 0°C to 100°C
(c) 4°C to 100°C (d) 4°C to 90°C

**25.** Percentage of water in our body weight is

(a) 79% (b) 50%
(c) 80% (d) 89%

**26.** Air contains large percentage of

(a) Water (b) Oxygen
(c) Nitrogen (d) Carbon dioxide

**27.** Example of change of state are

(a) melting (b) rusting
(c) evaporation (d) floating

**28. Match the items given in the list B with the items given in list A.**

| List A | List B |
|---|---|
| (1) Water is a universal _______ | (a) ice |
| (2) Water is composed of hydrogen and ________ | (b) 3% |
| (3) Liquid water is heavier than solid ________ | (c) water vapour |
| (4) Sea water is highly ________ | (d) atmosphere |
| (5) Water has maximum density at ________ | (e) water cycle |
| (6) Distilled water and rain water are the ________ forms of water. | (f) gas (vapour) |
| (7) Water stored under the soil is called ________ | (g) pure form |
| (8) Gaseous form of water is called ________ | (h) humidity |
| (9) The water suitable for drinking by human beings is called ________ water. | (i) dissolved |
| (10) Water exists in three forms solid (ice), liquid and ______ | (j) constant |
| (11) The moisture content in the atmosphere is called ______. | (k) weight |
| (12) The total water of the world remains ________ | (*l*) under- ground water |
| (13) A fish survives in water because it can breath the air _______ in water | (m) saline |
| (14) The earth is surrounded by a layer of air called the ______ | (n) potable |
| (15) Air occupies space and has ______ | (o) 4°c |
| (16) Water available for human consumption on the surface of the earth | (p) purest |
| (17) Water in the ocean is saline (salty) but the large amount of water evaporated from its coming down as rain water is not salty because it is in | (q) oxygen |
| (18) The process by which water from the surface of the earth comes back to earth by rain is known as | (r) solvent |

**29. Solve the given crossword with the help of the following clues.**

**ACROSS**

**3.** Gas formed by burning of carbon and breathed out by animals in respiration.

**5.** The process of giving off watery vapours from the surface of the pores of leaves.

**9.** Water suitable for drinking by human beings.

**11.** The change of a liquid into a vapour at a temperature below the boiling point.

**15.** The change of a gas or vapour to a liquid.

**DOWN**

**1.** Fog polluted by smoke

**2.** Collection and storing of rain water for future use.

**4.** A colourless gas forming about 80% of the atmosphere.

**6.** The process of removing salt from sea water.

**7.** The continuous process by which water is distributed throughout the Earth starting from evaporation of water from oceans and the land to the atmosphere and then back to the earth.

**8.** Continuous dry weather.

**10.** The process by which green plants use sunlight to convert carbon dioxide taken from air and water into carbohydrates.

**12.** A gas existing in air and combining with hydrogen to form water. Required for combustion and essential for life.

**13.** The mixture of gases that surrounds the Earth or some other celestial bodies.

**14.** A current of air, especially a natural one that moves along or parallel to the ground.

## Answers

**1.** (d) **2.** (d)

**3.** (a) condensation (b) evaporation
(c) condensation (d) evaporation
(e) evaporation

**4.** (a) False (b) False (c) True
(d) False (e) True

**5.** (a)

**6.** (1) Water vapour in air cools (2) clouds
(3) rain (4) liquid water
(5) ocean, lakes, and rivers

**7.** (c) **8.** (a) **9.** (d) **10.** (a)
**11.** (d) **12.** (d) **13.** (b) **14.** (d)
**15.** (b)
**16.** (i) condensation (ii) precipitation (iii) evaporation
**17.** (b) **18.** (c) **19.** (b) **20.** (a)
**21.** (d) **22.** (i) melted, (ii) frozen

**23.** (c) **24.** (d) **25.** (d) **26.** (c)
**27.** (a) and (c)
**28.** (1) – (r) (2) – (q) (3) – (a) (4) – (m) (5) – (o) (6) – (p) (7) – (*l*) (8) – (c) (9) – (n) (10) – (f) (11) – (h) (12) – (j) (13) – (i) (14) – (d) (15) – (k) (16) – (b) (17) – (g) (18) – (e)

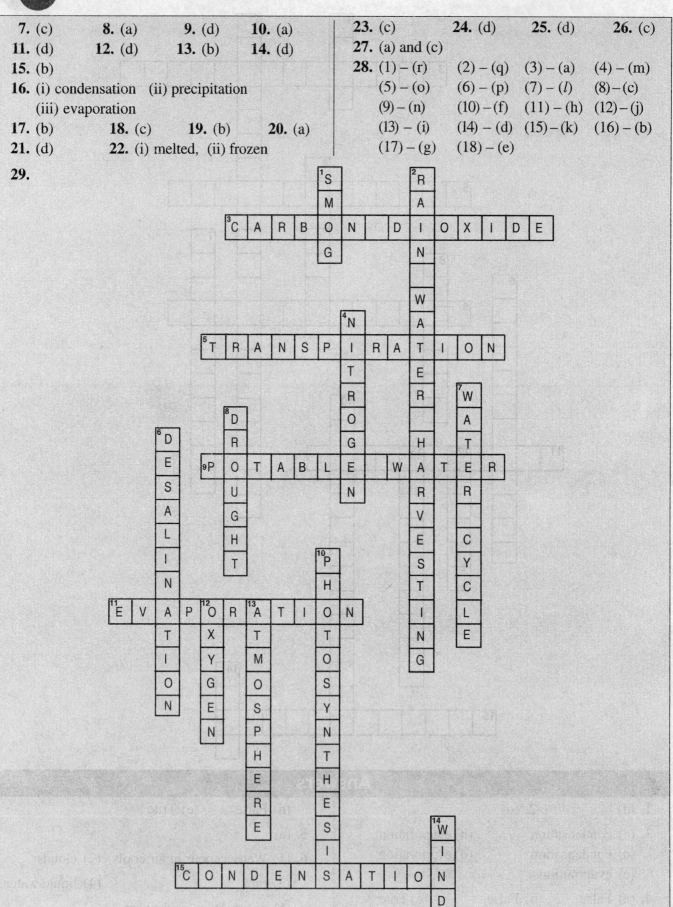

**29.**

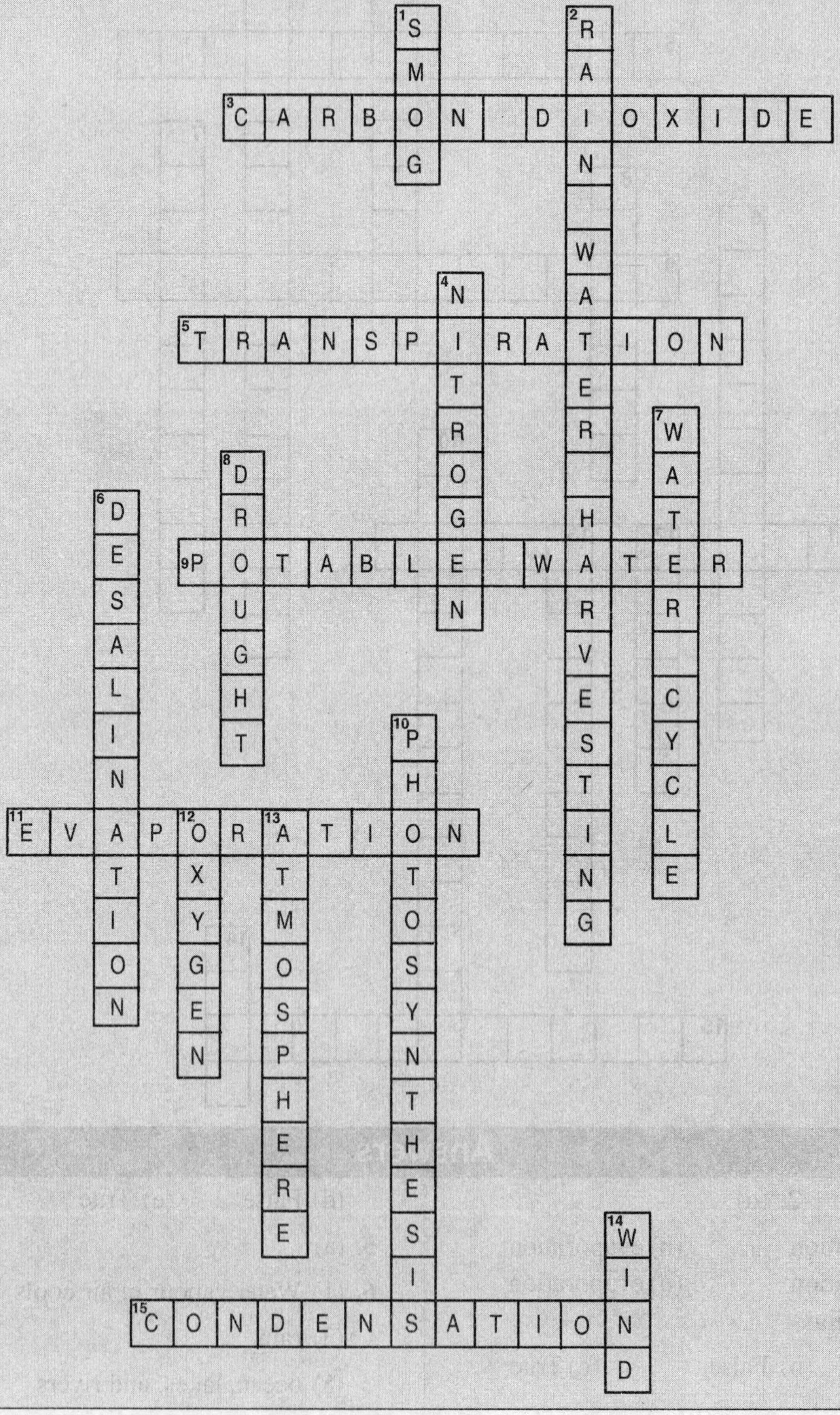

## Self Assessment Sheet-10

**1.** Sarita stored some blue crystals in water. She obtained a clear, blue liquid. Which of the following is correct about the change that took place ?

(a) The crystals melted and spread out in water to give a solution.

(b) The crystals dissolved in water to give a solution.

(c) The crystals were absorbed by water to give a suspension.

(d) The crystals dissolved in water to give a suspension.

**2.** In the given experiment, some sea water was heated in a beaker. After some time, it was observed that the number of water droplets formed on the aluminium sheet become less and less. Why was this so ?

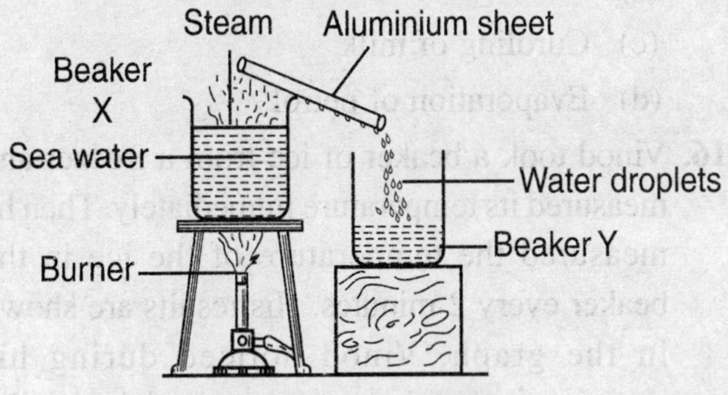


(i) The aluminium sheet had become hotter

(ii) The rate of condensation had decreased

(iii) Beaker X had become hotter

(a) (i) and (iii) only

(b) (i) and (ii) only

(c) (ii) and (iii) only

(c) (i), (ii) and (iii)

**3. Match the statements in the column I with those in column II.**

| *Column I* | *Column II* |
|---|---|
| (i) The process of removing insoluble particles of suspension by passing it through filter paper. | a. Sieving |
| (ii) The process of obtaining soluble salt from the solution. | b. Churning |
| (iii) The process of obtaining butter from curd. | c. Loading |
| (iv) The process of making particles of mud heavy with alum. | d. Filtration |
| (v) The process of separation of chaff from wheat flour. | e. Evaporation |

**(a)** (i) – e, (ii) – b, (iii) – a, (iv) – c, (v) – d

**(b)** (i) – d, (ii) – e, (iii) – b, (iv) – c, (v) – a

**(c)** (i) – e, (ii) – a, (iii) – b, (iv) – c. (v) – d

**(d)** (i) – b, (ii) – e, (iii) – c, (iv) – a, (v) – d

**4.** Which of the following mixtures will form a clear solution ?

(a) Water in castor oil

(b) Water in kerosene

(c) Salt in castor oil

(d) Salt in water

**5.** Which of the following objects cannot be made from plastic?

(a) Buckets (b) Frying pans

(c) Plates (d) Water bottles

**6.** Which of the following represents condensation?

(a) Drying of wet clothes

(b) Melting of glacial ice

(c) Hardening of molten chocolate in a referigerator

(d) Accumulation of water droplets on a cold bottle.

**7.** Which of the following methods is not involved in the process of separation of grains from stalks, husks, chaff and bran?

(a) Sieving (b) Handpicking

(c) Winnowing (d) Threshing

**8.** Which of the following properties cannot be used to distinguish between a solid and a liquid?

(a) Solubility (b) Compressibility

(c) Change of state (d) Conductivity

**9.** Shubham wanted to find out whether salt dissolved better in tap water or in rain water. He added salt to each beaker shown below until no more salt could be dissolved. State two ways in which Shubham's experiment was not a fair one.

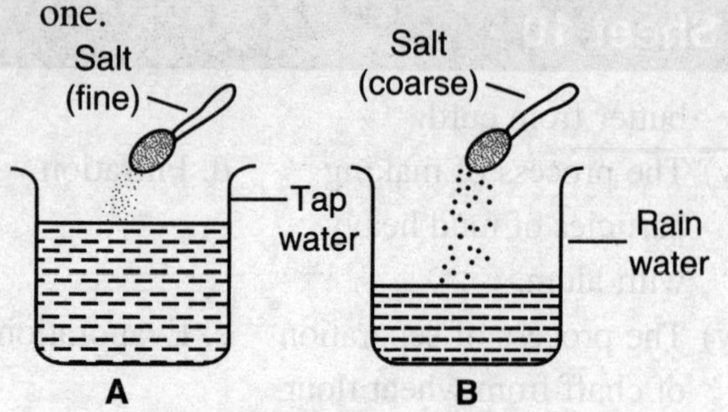


**10.** The water cycle in nature involves the following processess sequentially :
(a) evaporation, precipitation, condensation
(b) precipitation, condensation, evaporation
(c) evaporation, condensation, precipitation
(d) condensation, evaporation, precipitation

**11.** Which of the following is an endothermic reaction?
(a) Formation of curd from milk
(b) Production of gas when quicklime is added to water.
(c) Opening of a soda bottle with production of gas
(d) Evaporation of water

**12.** If you are given a mixture of sand and salt and water. You can separate these by the process of
(a) only decantation (b) only filtration
(c) only evaporation (d) only condensation
(e) using all of the above.

**13.** In an experiment :
(i) Vinegar was added into a bottle containing a table spoon of baking soda in it.
(ii) A balloon was now put at the mouth of the bottle, the balloon got inflated due to the production of carbon dioxide gas in the bottle.

This is an example of :
(a) physical and reversible change
(b) physical but irreversible change
(c) chemical and reversible change
(d) chemical but irreversible change

**14.** Which of the following are desirable changes ?
(i) Drying of clothes
(ii) Melting of ice cream
(iii) Cement getting hard when exposed to moisture
(iv) Desalination of sea water
(a) (i) and (ii) only (b) (i) and (iv) only
(c) (ii) and (iii) only (d) (ii) and (iv) only

**15.** Which of the following changes is slow and periodic?
(a) Rotation of the earth
(b) Rusting of iron
(c) Curdling of milk
(d) Evaporation of petrol

**16.** Vinod took a beaker of ice from a freezer and measured its temperature immediately. Then he measured the temperature of the ice in the beaker every 2 minutes. His results are shown in the graph. Vinod noticed during his experiment that the temperature of the ice did not change for a period of time.

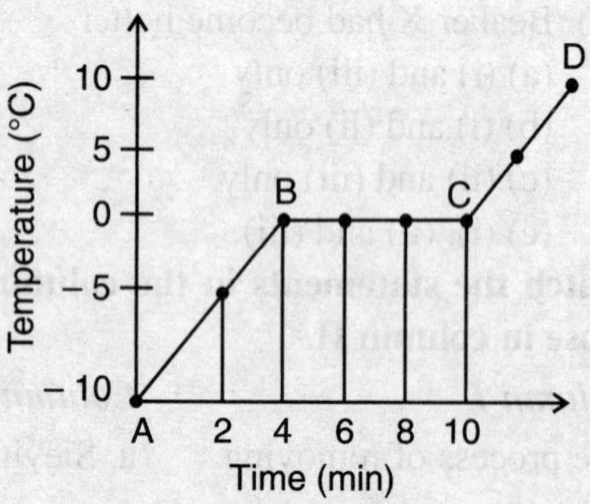


(a) Which part of the graph corresponds to this period of time?
(b) How long was this period of time ?
(c) What change did the ice undergo during this period of time?

**17.** Complete the diagram below by filling in the processes that take place in 2, 3 and 4 respectively.

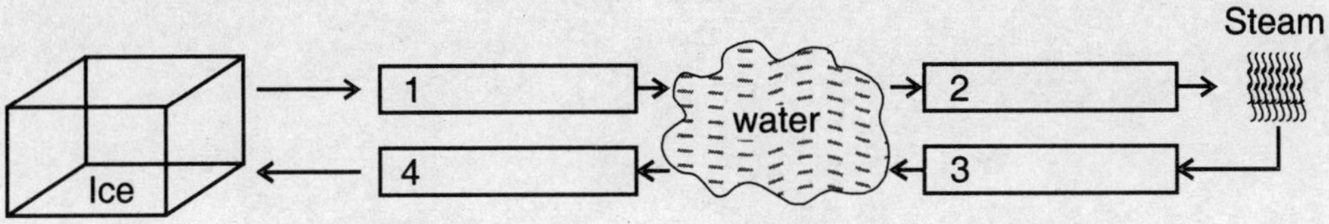


**18.** How can you separate the gradients of the following mixtures ?

(a) Chalk powder added to water and stirred.

(b) Cooking oil added to sauce and stirred.

**19.** Cooking of an egg in boiling water is a

(a) Permanent change

(b) Temporary change

(c) Reversible change

(d) Both (b) and (c)

**20. Match correctly.**

| List A | List B |
|---|---|
| (i) Pure water has maximum volume at | (a) ice |
| (ii) Water is a universal | (b) increases |
| (iii) Sea water is highly | (c) 0°C |
| (iv) Water has maximum density at | (d) solvent |
| (v) Liquid water is heavier than solid | (e) saline |
| (vi) When water at 4°C is cooled or heated, its volume | (f) 4° C |

## Answers

**1.** (b) **2.**(b) **3.** (B) **4.** (d)

**5.** (b) Plastic is a bad conductor of heat and does not allow heat to pass through it

**6.** (d) **7.** (b) **8.** (c)

**9.** He used different types of salt and different amounts of water in A and B.

**10.** (c) **11.** (a) **12.** (e) **13.** (d)

**14.** (b) **15.** (a)

**16.** (a) BC (b) 6 minutes (c) It melted.

**17.** (1) melting (2) boiling (3) condensation (4) freezing

**18.** (a) Filtration with fine filter paper

(b) Scoop off floating oil.

**19.** (a)

**20.** (i) – (c), (ii) – (d), (iii) –(e), (iv) –(f), (v) – (a), (vi) – (b)